Aluminium Design and Construction

Aluminium Design and Construction

John Dwight

MSc, FI Struct E
Former Reader in Structural Engineering,
University of Cambridge; and Fellow of Magdalene College, Cambridge

CRC Press
Taylor & Francis Group
Boca Raton London New York

CRC Press is an imprint of the
Taylor & Francis Group, an **informa** business

CRC Press
Taylor & Francis Group
6000 Broken Sound Parkway NW, Suite 300
Boca Raton, FL 33487-2742

First issued in paperback 2019

Typeset in 10/12 Palatino by
Mathematical Composition Setters Ltd, Salisbury, Wiltshire

ISBN-13: 978-0-419-15710-6 (hbk)
ISBN-13: 978-0-367-86629-7 (pbk)

British Library Cataloguing in Publication Data
A catalogue record for this book is available from the British Library

Library of Congress Cataloging in Publication Data
Dwight, J. B. (John B.), 1921–
Aluminium design and construction / J.B. Dwight.
p. cm.
Includes bibliographical references and index.
ISBN 0-419-15710-7
1. Aluminum construction. 2. Aluminum. 3. Aluminum, Structural.
4. Structural design — Standards — Europe. I. Title
TA690.D855 1998
624.1'826 — dc21 98-39235
CIP

Visit the Taylor & Francis Web site at
http://www.taylorandfrancis.com

and the CRC Press Web site at
http://www.crcpress.com

Contents

Preface

Aluminium is easily the second most important structural metal, yet few designers seem to know much about it. Since the 1940s, as aluminium rapidly became more important, engineers have been slow to investigate what it has to offer and how to design with it. Aluminium is hardly mentioned in university courses. This book is a contribution to an educational process that still seems to be needed.

The object of this book is to provide a conversion course for engineers already familiar with steel. In fact, structural aluminium, a strong ductile metal, has much similarity to steel and design procedures are not very different. Chapters 1–4 give general information about aluminium and aluminium products, Chapter 4, with its coverage of the thorny subject of the alloys, being particularly important. The rest of the book (Chapters 5–12) provides rules for making structural calculations and the reasoning that lies behind them. The treatment is mainly aimed at the construction industry.

Weight saving is more important in aluminium than in steel, because of the higher metal cost. More accurate design calculations are therefore called for. Critical areas in aluminium include buckling, deflection, weld strength and fatigue. Other aspects which do not arise at all in steel are the use of extruded sections, heat-affected zone (HAZ) softening at welds and adhesive bonding. This book covers these fully.

The aim had been to follow the design rules in British Standard BS.8118 (Structural Use of Aluminium), one of the first codes to be written in limit state format, and much of the book in fact does this. However, there are some areas where the writer feels that the British Standards approach is other than ideal, and for these the book provides alternative rules which are simpler, more correct or more economical. Such areas include limiting design stresses, HAZ softening, local buckling and asymmetric bending. A further feature is the inclusion of Chapter 10, which explains how to obtain the section properties of complex extruded shapes, including torsional properties.

At the time of writing (1998) a draft version has appeared of the new aluminium Eurocode (EC9), which will in time supersede the various national codes. This document is referred to in the book.

Acknowledgements

I am indebted to the following people who have provided unstinting help:

Michael Bayley, British Alcan;
Dr P.S. Bulson CBE, former head of MEXE;
Ron Cobden, aluminium designer, formerly with British Alcan;
Professor S.L. Harris, formerly of Lancaster University;
Dr G.H. Little, Birmingham University;
Richard Mahoney and David Keevil, Aluminium Federation;
Professor C.D. Marsh, Concordia University, Montreal, who has been designing in aluminium for 50 years and still is;
Dr O.T. Midling, Hydro Aluminium, Norway;
Dr N.S. Moss and Dr J. Powell, CIBA;
Professor D.A. Nethercot, Nottingham University;
Dr M.H. Ogle, TWI;
Dr Ian Robertson, Cegelec (Alstom), France;
Morris de Rohan, Agent General for South Australia, London;
Professor F. Soetens, TNO Bouw, Delft;
Wayne Thomas, TWI;
Philip Tindall, Hyder Consulting;
Professor N.S. Trahair, University of Sydney;
Don Webber, MEXE;
Dr Roy Woodward, formerly with Aluminium Laboratories, Banbury.

I would also like to thank Marica de Lopez and Susan Bennett who have done a noble job in typing the text.

Finally, I reserve extra special thanks for my wife Jo, who has cheerfully put up with aluminium pervading the house for far too long.

List of symbols

The following symbols appear generally thoughout the book. Others are defined as they arise.

A	area of section
A_e	area of effective section
A_w	area of weld deposit
A_z	area of nominal heat-affected zone (HAZ)
D	overall depth of section
E	modulus of elasticity
G	shear modulus
H	warping factor
I	second moment of area ('inertia')
I_p	polar inertia
I_{uu}, I_{vv}	principal inertias
I_{xx} (I_{yy})	inertia about centroidal axis xx (yy)
I_{xy}	product of inertia
$\mathfrak{J}$	St Venant torsion factor
L	unsupported member length for overall buckling
M	moment arising under factored loading
M_c	calculated moment resistance
$\bar{M}$	moment arising per unit length of weld under factored loading
$\bar{M}_c$	calculated moment resistance per unit length of weld
N	fatigue endurance (cycles to failure)
P	axial force arising under factored loading
P_c	calculated axial force resistance
$\bar{P}$	force arising under factored loading, per fastener or per unit length of weld
$\bar{P}_c$	calculated force resistance, per fastener or per unit length of weld
$\bar{R}$	reaction force between plates per bolt (friction-grip bolting)
S	plastic section modulus (uniaxial moment)
S_m	plastic section modulus (biaxial moment)
S_p	reduced plastic modulus in presence of axial load (uniaxial moment)

S_{pm}	reduced plastic modulus in presence of axial load (biaxial moment)
T	torque
T	bolt tension (friction-grip bolting)
T_0	proof load for HSFG bolt
T_0	temperature of metal prior to welding
T_1	external force per bolt (friction-grip bolting)
V	shear force arising under factored loading
V_c	calculated shear force resistance
W	applied load
XX, YY	convenient axes, not through centroid
Z	elastic section modulus
a	spacing of transverse stiffeners
b	width of plate element
b_e	effective width of plate element
c	depth of lip reinforcement on an outstand element
c	imperfection factor in overall buckling
d	depth of web element
d	hole diameter
d_c, d_t	depth of web in compression and tension
e	strain
el	percentage elongation
f	stress arising under nominal loading (fatigue)
f_o	minimum 0.2% proof stress
f_r	stress-range (fatigue)
f_u	minimum tensile strength
g	slenderness adjustment factor for plate element, under strain gradient
g	weld throat dimension
g	distance of shear centre S from centroid G
h	height of load application point above centroid (lateral-torsional buckling)
h	weld fusion boundary dimension
k_z	heat-affected zone (HAZ) softening factor
l	effective length for overall buckling
mm	axis about which applied moment acts (biaxial bending)
nn	neutral axis (biaxial bending)
p_a, p_b, p_o, p_v	limiting stresses for members design (Table 5.2)
p_f, p_p, p_s, p_t, p_w	limiting stresses for joint design (Table 5.2)
p_v	limiting shear stress in adhesive
q_1	shear stress arising in adhesive under factored loading
r	radius of gyration
s	heat-affected zone (HAZ) dimension
ss	axis of symmetry
t	metal thickness

uu, *vv*	principal axes
w	effective size of weld
w	unit warping (torsion)
xx, *yy*	centroidal axes
y	distance from *xx*
y_E	distance of element centroid from *xx*
z	longitudinal warping movement (torsion)
α	oversize factor for small welds
α_1, α_0	effective width factors (local buckling)
β	slenderness parameter for plate element (local buckling)
β_f	limiting value of β for fully compact section
β_s	limiting value of β for semi-compact section
β_x	special lateral-torsional buckling factor
γ_f	limit state of static strength, factor applied to loads ('loading factor')
γ_m	limit state of static strength, factor applied to resistance ('material factor')
γ_s	serviceability factor (friction-grip bolting)
Δ	modified slenderness parameter for plate elements
Δ	beam deflection
ε	non-dimensionalizing factor = $\sqrt{(250/p_o)}$
λ	slenderness parameter for overall buckling
μ	slip-factor (friction-grip bolting)
ν	Poisson's ratio
ρ_1, ρ_0	adjustment factors for reinforced plate elements
σ	stress
σ_o	0.2% proof stress
σ_{cr}	elastic critical stress (plate elements)
σ_m	mean stress at failure (plate elements)
τ	shear strength of adhesive
ψ	strain-gradient parameter (plate elements)

Symbols defining units (conversion factors):

Length	m	metres	(1 m = 3.2808 ft)
	mm	millimetres	(1 mm = 0.03937 in)
Force	N	newtons	(1N = 0.2250 lbf)
Stress	N/mm^2	MPa megapascals	(1N/mm^2 = 145.0 psi)

CHAPTER 1

About aluminium

1.1 GENERAL DESCRIPTION

1.1.1 The element

Aluminium is a metallic element having the chemical symbol Al, with atomic number 13 and atomic weight 27. The nucleus of the atom contains 13 protons and 14 neutrons (a total of 81 quarks). Aluminium is the third most common element in the earth's crust, coming after oxygen and silicon. It makes up 8% of the crust's total mass and is the most abundant metal.

1.1.2 The name

Since birth it has been dogged with a long inconvenient name (actually from the prenatal stage). And it suffers in having two different versions in common use: the N. American *aluminum* and the European *aluminium*. The name was coined by Sir Humphry Davy in about 1807 (based on a Latin word *alumen*), although at that stage the element did not actually exist in metallic form. Davy's proposal was the shorter word (aluminum), but by the time commercial production began in the 1850s the extra 'i' had crept in. The two versions have co-existed to this day.

One wonders why the industry has done nothing to replace its four or five-syllable encumbrance with a simple user-friendly name, like the monosyllables enjoyed by other common metals. A step in the right direction would be to adopt the N. American version ('aloominum') worldwide, since this takes half as long to say as the European one. Better still would be to move to a snappier word altogether. The abbreviation 'alli' is often used in speech; why not adopt this as the official name, or even just 'al'? Charles Dickens expressed just such a sentiment back in 1856 when he wrote:

> Aluminum or as some write it, Aluminium, is neither French nor English; but a fossilised part of Latin speech, about as suited to the mouths of the populace as an icthyosaurus cutlet or a dinornis marrow-bone. It must adopt some short and vernacular title.

1.1.3 The industrial metal

It is only since 1886 that aluminium has been a serious industrial metal, that being the year when the modern smelting process was invented. It has thus been available for a very short time, compared to the thousands of years that we have had bronze, copper, lead, iron, etc. Today it easily leads the non-ferrous metals in volume usage. It is selected in preference to steel in those areas where its special properties ('light and bright') make it worth paying for. There are many applications where aluminium has found its rightful niche, and others where it is still on the way in. Current world consumption is some 20 million tonnes per year.

1.1.4 Alloys

Pure aluminium is weak, with a tensile strength ranging from about 90 to 140 N/mm^2 depending on the temper. It is employed for electrical conductors and for domestic products (such as pans, cans, packaging), but for serious structural use it has to be strengthened by alloying. The strongest alloys have a tensile strength of over 500 N/mm^2.

There are around ten basic alloys in which wrought material (plate, sheet, sections) is produced. Unfortunately, each of these alloys appears in a vast range of different versions, so that the full list of actual specifications is long. The newcomer therefore finds material selection less simple that it is in structural steel, and there is also the alloy numbering system to contend with.

In engineering parlance the term *aluminium* (or *aluminum*) covers any aluminium-based material, and embraces the alloys as well as the pure metal. To refer specifically to the pure or commercially pure material, one has to say 'pure aluminium'.

1.1.5 Castings

Aluminium is eminently suitable for casting. For larger items (such as sand-castings), aluminium is often a preferred option to cast iron. For smaller items (such as dye-castings), it provides a strong alternative to zinc. A wide range of casting alloys is available, different from the wrought alloys. The reliability that is possible with aluminium castings is demonstrated by the fact that they have become standard for car wheels.

1.1.6 Supposed health risk

For many years it was believed that aluminium was entirely non-toxic, and superior in this respect to most other metals. In the 1980s this picture was reversed when researchers claimed to show that the prolonged use of aluminium saucepans could cause minute amounts of the metal to be absorbed in the brain, and that this in turn could increase the risk of Alzheimer's disease (a type of senile dementia). Following coverage in the media, aluminium cookware quickly went out of fashion and many people became fearful of aluminium in any form.

Later work has cast serious doubt on this finding, and modern opinion is that the 'aluminium theory' for the increased risk of Alzheimer's disease was grossly overstated. There are two questions. First, do we really absorb significant quantities of aluminium from saucepans, relative to what we take in anyway from our food and drink? Second, at what level of absorption does aluminium become harmful? The answer to the first question is: No, because of the tenacious oxide film always present on aluminium. The answer to the second is: Nobody knows. In any case there is no need to fear aluminium outside the culinary scene.

1.1.7. Supposed fire risk

Three of the British warships sunk in the Falklands war of 1981 had aluminium superstructures. At the time, the press stated that in the severe fires preceding these sinkings the aluminium had actually burnt. This was completely untrue. The aluminium structures lost strength and distorted, but did not burn. Aluminium sections, plate, sheet, foil and wire will not support combustion. Only in the form of very finely divided powder or flake can the metal be made to burn, as can finely divided steel. Magnesium is a different story.

1.2 PHYSICAL PROPERTIES

Below we quote the physical properties of aluminium that are of interest in design, the values for the alloys being close to those for the pure metal in most cases. More precise values appear in *The Properties of Aluminium and its Alloys* issued annually by the Aluminium Federation in Birmingham, UK [1].

(a) Weight

The density ρ of pure aluminium compares as follows with steel:

Pure aluminium	$\rho = 2.70\ g/cm^3$
Structural steel	$\rho = 7.9\ g/cm^3$

and the value for the alloys used for wrought products lies in the range 2.67–2.80 g/cm^3. A rounded value of 2.7 g/cm^3 is normally used in design, leading to the following practical formulae:

	Mass	*Weight*
Sections	$0.0027A$ kg/m	$0.027A$ N/m
Plate, sheet	$2.7t$ kg/m^2	$27t$ N/m^2

where A is the section area (mm^2) and t the plate thickness (mm). Table 1.1 compares aluminium with other metals.

(b) Elastic constants

Aluminium is a springy metal with a relatively low modulus of elasticity (E). For the pure metal at room temperature it compares with steel as follows:

Pure aluminium	$E = 69$ kN/mm^2
Structural steel	$E = 205$ kN/mm^2

while the value for wrought alloys lies in the range 69–72 kN/mm. For design purposes British Standard BS.8118 adopts a standard figure of $E = 70$ kN/mm^2, which is similar to that for glass. For aluminium E decreases steadily with temperature, dropping to 67 kN/mm^2 at 100°C and 59 kN/mm^2 at 200°C.

Poisson's ratio (ν) is higher than the accepted figure of 0.30 used for steel and should be taken equal to 0.33, based on the work of Baker and Roderick at Cambridge in 1948 [2].

The corresponding figure for the shear modulus (G), based on the above values of E and ν, is:

$$G = \frac{E}{2(1+\nu)} \approx 26 \text{ kN/mm}^2$$

(c) Thermal expansion

The coefficient of linear expansion α for pure aluminium at room

Table 1.1 Density of metallic elements

Element		Density (g/cm^3)	Density relative to Al
Lithium	Li	0.53	0.20
Magnesium	Mg	1.74	0.64
Aluminium	Al	2.70	1.00
Titanium	Ti	4.51	1.67
Zinc	Zn	7.13	2.64
Tin	Sn	7.28	2.70
Iron	Fe	7.87	2.92
Copper	Cu	8.93	3.31
Silver	Ag	10.50	3.89
Lead	Pb	11.34	4.20
Gold	Au	19.28	7.15
Osmium	Os	22.58	8.37

temperature compares with steel as follows:

Pure aluminium	$\alpha = 23.5 \times 10^{-6}/°C$
Structural steel	$\alpha = 12 \times 10^{-6}/°C$

with the value for the wrought alloys lying in the range $22–24.5 \times 10^{-6}$. British Standard BS.8118 adopts a rounded value of $23 \times 10^{-6}/°C$ for use in design. Note that α increases steadily with temperature, going up to 26×10^{-6} at 200°C.

(d) Melting point

This is 660°C for pure aluminium compared with 1500°C for mild steel, while the values for the alloys are somewhat lower. The boiling point is 1800°C.

(e) Thermal constants

Aluminium is a valid material for use in heat exchangers as an alternative to copper, the thermal conductivity of the pure metal at room temperature being 240 W/m°C (about four times the figure for steel). However, the conductivity is drastically reduced by alloying, down 50% for some alloys. The specific heat of pure aluminium at room temperature is 22 cal/g°C (about twice the steel value).

(f) Electrical conductivity

Pure aluminium competes with copper in some electrical applications, and is the standard material for the conductors in overhead transmission lines (with a steel strand up the middle). The resistivity of very pure aluminium is 2.7 $\mu\Omega$ cm at room temperature. Again the value is highly sensitive to alloying and is twice the above value for some alloys.

1.3 COMPARISON WITH STEEL

Wrought aluminium in its alloyed form is a strong ductile metal and has much similarity to structural steel. Its mechanical properties tend to be inferior to those of steel, the stronger alloys being comparable in strength but less ductile. The approach to structural design is much the same for the two metals and below we concentrate on the differences. Unlike steel, aluminium is, of course, non-magnetic.

1.3.1. The good points about aluminium

Lightness

Aluminium is light, one third the weight of steel.

Non-rusting

Aluminium does not rust and can normally be used unpainted. However, the strongest alloys will corrode in some hostile environments and may need protection.

Extrusion process

This technique, the standard way of producing aluminium sections, is vastly more versatile than the rolling procedures in steel. It is a major feature in aluminium design.

Weldability

Most of the alloys can be arc welded as readily as steel, using gas-shielded processes. Welding speeds are faster.

Machinability

Milling can be an economic fabrication technique for aluminium, because of the high metal removal rates that are possible.

Glueing

The use of adhesive bonding is well established as a valid method for making structural joints in aluminium.

Low-temperature performance

Aluminium is eminently suitable for cryogenic applications, because it is not prone to brittle fracture at low temperature in the way that steel is. Its mechanical properties steadily improve as the temperature goes down.

1.3.2 The bad points

Cost

The metal cost for aluminium (sections, sheet, plate) is typically about 1.5 times that for structural steel volume for volume. For aircraft grade material, the differential is much more. However, fabrication costs are lower because of easier handling, use of clever extrusions, easier cutting or machining, no painting, simpler erection. Thus, in terms of total cost, the effect of switching to aluminium is usually much less than one would expect. An aluminium design can even be cheaper than a steel one.

The other side of the cost picture is aluminium's relatively high scrap value, a significant factor in material selection for components designed to have a limited life. But high scrap value may not always be a good thing, as it encourages the unscrupulous to come on a dark night and remove an aluminium structure for sale as scrap. This has happened.

Buckling

Because of the lower modulus, the failure load for an aluminium component due to buckling is lower than for a steel one of the same slenderness.

Effect of temperature

Aluminium weakens more quickly than steel with increasing temperature. Some alloys begin to lose strength when operating above 100°C.

HAZ softening at welds

There tends to be a serious local drop in strength in the heat-affected zone (HAZ) at welded joints in some alloys.

Fatigue

Aluminium components are more prone to failure by fatigue than are steel ones.

Thermal expansion

Aluminium expands and contracts with temperature twice as much as steel. However, because of the lower modulus, temperature stresses in a restrained member are only two-thirds those in steel.

Electrolytic corrosion

Serious corrosion of the aluminium may occur at joints with other metals, unless correct precautions are taken. This can apply even when using alloys that are otherwise highly durable.

Deflection

Because of the lower modulus, elastic deflection becomes more of a factor than it is in steel. This is often a consideration in beam design.

1.4 HISTORY

The following is an outline of how, in the century from 1845 to 1945, aluminium became the second most important industrial metal [3].

1.4.1 The precious metal stage

In 1825 the Danish scientist Hans Oersted succeeded in isolating minute particles of metallic aluminium. Having satisfied himself that the metal existed he lost interest, and returned to his main field of study, electromagnetism. Two years later Friedrich Wöhler, professor of chemistry at Göttingen University in Germany, began to repeat Oersted's experiments. He persisted for 20 years and eventually produced the first actual nuggets of aluminium, big enough to reveal some of the metal's special properties, especially its lightness and brightness. A piece of the aluminium he produced is on display in the university museum at Göttingen.

The step from scientific curiosity to commercial product was made in the 1850s by the French chemist Henri Ste-Claire Deville, a professor at the Sorbonne University in Paris. Having read of Wöhler's work, he interested Napoleon III in the new metal and received funding to develop it commercially. Wöhler's chemical process for obtaining aluminium had relied on potassium, the high cost of which meant that it would never be of commercial interest. Ste-Claire Deville's approach was to develop an economic method for obtaining sodium, and use this instead of potassium in the Wöhler process. He was thus able to produce aluminium which, if hardly cheap, would be a practical metal for some applications. A dozen small bars were displayed at an exposition in Paris in 1855, dubbed 'silver from clay'. The first aluminium artefact is said to have been a rattle made for the infant son of Napoleon III.

Over the next 30 years the Deville process was exploited in several countries, including Britain, to produce what was at that stage a high cost luxury metal. An example of its use was for spoons at royal banquets instead of gold, and it was also employed in cast form for small statues. A purity of about 97% is recorded for metal produced at the Nanterres factory outside Paris.

1.4.2 The big breakthrough

The breakthrough came in 1886 when two young men, aged 22, independently invented the same electrolytic process for smelting aluminium. One was the American scientist Charles M. Hall, and the other Paul Héroult of France. Hall got his patent in two months before Héroult, but they should be given equal credit for their great achievement.

Electrolysis had been tried before by Humphry Davy (in the 1800s) and by Robert Bunsen (in the 1850s) without success. Héroult and Hall got the process to work. Their method used alumina as the raw material and molten cryolite for the electrolyte, together with a lot of electricity. It was made possible by the recent advent of cheap electric power. A further essential step, two years later, was the development by Karl Bayer of Germany of an economic process for obtaining alumina from the ore bauxite. The Hall–Héroult process, using alumina from the Bayer process, slashed the price of aluminium by a factor of ten and turned it into a tonnage industrial metal. These are the processes still used.

1.4.3 Early applications

In the years following 1886, the newly available metal was tried for a wide range of uses, some successful, some less so. A famous example from that time is the statue of Eros in Piccadilly Circus, London, cast in 1893. Another is the sheet metal roofing of the dome of the church of San Gioacchino in Rome in 1897. Both of these are still in good condition.

There were various attempts to develop aluminium as a structural material, some of which were abandoned, only to be successfully resurrected 50 years later. Aluminium was used for panelling on railway coaches, and Russia placed an order for aluminium goods-wagons. In 1986 Yarrow's of Scotland employed aluminium plating for a torpedo-boat. Some early motor-cars had aluminium in their bodywork. There is a report of a demountable boat to be carried into the interior of Africa by European explorers (actually by their porters). And another of a prefabricated portable aluminium house invented by Mr Howes of Seattle, USA, weighing 70 kg, for use by gold-diggers in the Klondyke (1897). Unfortunately we know little as to how those early efforts fared, except that Yarrow's torpedo-boat was a failure, because it was in an unsuitable 6% Cu alloy which corroded.

Turning to the aero scene, it is fitting that the engine of the world's first successful powered aircraft (flown by the Wright brothers in 1903) had cylinder-block and crankcase in aluminium. But this was not the metal's first appearance in an aeroplane. Two years earlier Wilhelm Kreiss had used aluminium for the floats on his triplane, in which he failed to get airborne from a reservoir in Austria.

1.4.4 Establishment of the alloys

In order for aluminium to be useful as a structural metal, it was essential to develop suitable alloys, since the pure metal was rather weak. The pioneer of alloy development was the German metallurgist Alfred Wilm, who discovered *age hardening* – the phenomenon whereby some alloys of aluminium, although still weak just after heat treatment (quenching), slowly harden when left for several days at room temperature. He first demonstrated this behaviour in 1903 with an alloy containing 4% Cu. Further work led to stronger alloys, and eventually in 1909 he produced one having properties nearly as good as mild steel, an Al–CuMgMn composition which he christened 'duralumin'. It was the start of what we now term the 2xxx-series alloy group.

A scientific explanation of age-hardening did not appear until 1920, soon after which a second kind of age-hardening alloy emerged, namely the Al–MgSi type. This alloy group (the present day 6xxx series) has a tensile strength in its strongest version of some 300 N/mm^2, and is thus generally weaker than the 2xxx series. But it has other features that have since led it to become the 'mild steel' of aluminium.

Next to appear were the Al–Mg alloys, our present 5xxx series. These are non-heat-treatable alloys that are hardened by rolling, developed in Britain in the late 1920s and marketed under the name 'Birmabright'. These alloys have extra good corrosion resistance and during the 1930s were successfully used in boat building.

The 1930s also saw the arrival of the two strongest alloy types, both of which are heat-treatable materials that respond to *artificial ageing*, i.e. enhanced ageing at an elevated temperature. The first, another 2xxx-series alloy, was a development of Wilm's duralumin and was christened 'superdural'. It has a tensile strength of over 450 N/mm^2 when fully heat-treated and a high elastic limit. The second, appearing in 1936, was stronger still. This was a new Al–Zn Mg type of alloy, the first of our present day 7xxx series, having a tensile strength of over 500 N/mm^2 and still the strongest form of aluminium that is generally used.

By 1939 all of today's main alloys had thus arrived except one, namely the weldable kind of 7xxx-series alloy. This was actively developed after the war.

1.4.5 The first major market

The application which put aluminium on the map as a structural metal was its use in aircraft, first in airships and later in aeroplanes [4].

The airship pioneer was the German general Count Ferdinand von Zeppelin, whom we may regard as the father of structural aluminium. His first 'zeppelin' (the LZ1), which made its maiden flight over Lake Constance, Germany, in 1900, was 130 m long by 11.5 m diameter. The frame was fabricated from small sections formed from strip. In the LZ1, these were in pure aluminium, but for later designs Zeppelin was able to lighten the structure by switching to Wilm's duralumin. A sample of 'W-section' from the zeppelin L32, shot down over London in 1917, is preserved in the Shuttleworth Collection near Bedford (a museum of classic aircraft). Aluminium framed airships continued into the 1930s, but were abandoned in 1937 after some spectacular disasters; hydrogen was too dangerous. The ill-fated British R101 was 230 m long, one of the longest aluminium structures ever built (1930).

Aluminium's huge step forward was its use for military aircraft in World War 2. During the 1920s and 1930s aircraft construction had progressed from the old 'string and ceiling wax' concept to the all-metal stressed-skin airframe. By 1939 sophisticated designs had been perfected, using the strongest alloys and advanced structural analysis. Typical construction was in the form of stiffened sheet using thousands of rivets per aircraft, and as the war developed such aircraft were produced in stupendous quantities. In Britain, 22 000 Spitfires were built between 1936 and 1947, and this was only one design from one country.

1.5 ALUMINIUM SINCE 1945

1.5.1 Growth in output

The world's annual output of aluminium in 1943 at its wartime peak was about two million tonnes (fourteen times the figure for 1933), all of which was going into warplanes. In 1945 that market dried up and the aluminum industry had quickly to find other outlets. This was not a great problem, because aluminium could be used as a replacement for conventional materials that were then in short supply, and it had become effectively cheaper than before the war. Aluminium obtained an immediate toe-hold in many new markets. Some markets were only temporary and disappeared when steel became plentiful again, but most of them stuck and quickly grew. At one stage the annual output of aluminium was doubling every seven years. Today production is ten times what it was in 1943 (some twenty million tonnes per year), a quarter of which goes into packaging. The volume consumption of steel is some 12 times greater.

Aircraft still provide a very important market, but a far greater tonnage goes into other uses. The following is an account of how aluminium has developed as a structural metal (outside the aero field), and a review of its present position.

1.5.2 New technology

Most of aluminium's essential technology was in place by 1939. The following are postwar developments of structural relevance.

Alloy usage

Prewar the most used structural alloys were the 2xxx series (duralumin, superdural). Since 1945 the 5xxx and 6xxx series have largely taken over, and the 2xxx series is mainly confined to aircraft. Another development has been the introduction of weldable 7xxx material.

Large rivets

Small rivets had been employed by the million in airframes. After 1945 large aluminium rivets were developed for use in civil engineering structures. After some decades of use, these gave way to welding.

Welding

The most important structural development has been in welding. Prewar the only available arc-welding technique was ordinary stick welding with flux-coated electrodes (manual metal arc). This was unsatisfactory for aluminium because the necessary fluxes were highly corrosive, and the arc characteristics made it difficult to achieve a smooth deposit. The solution lay in processes using an inert gas (argon or helium) to shield the arc. The first to appear was TIG, which employs a tungsten electrode and separate filler wire. Then came the semi-automatic MIG process, using a continuous-feed wire electrode (subsequently adapted to provide the CO_2 process for steel). Both were developed in USA around 1950. Since then these processes have become standard practice in aluminium construction, except for joints in the very strong alloys for which arc-welding is no good. Recently (in the early 1990s), we have seen the invention, by the TWI (formerly The Welding Institute) in Britain, of the promising new friction-stir process.

Extruded sections

The extrusion of aluminium was already a viable process by 1939. Since 1945 there has been a vast increase in the use of structural sections produced by this method, mainly using 6xxx-series alloys. The sections are typically of ingenious and intricate shape, with an emphasis on profiles that are very thin relative to their overall size. Hollow sections have grown in popularity, and the price differential between these and non-hollows has nearly vanished. An important development has been the increasing use of large presses, with section widths up to 0.8 m being now available.

Adhesive bonding

A range of strong adhesives exists, which are especially suitable for making glued structural joints in aluminium. In some applications this is a preferred alternative to welding and its use is likely to grow.

Milling

Another development has been the installation of massive milling machines, enabling large integrally stiffened panels to be produced by machining them from the solid out of thick plate. This development is mostly of interest for aircraft.

1.5.3 Structural engineering

The heavy structural use of aluminium was pioneered in 1931 in the USA when a drag-line crane, having a 46 m mainly aluminium boom, started work on the Mississippi embankments [5]. Thirty such booms went into service. Better known is the uprating of the 220 m long Smithfield Street Bridge in Pittsburgh, USA, achieved by redecking it in aluminium (1934). The structural sections for the new deck were produced by hot-rolling in steel-mill rolls, large extrusions being then unavailable. Another early application was in a rail-bridge over the Grasse River at Massena, NY, USA, where one of the 26 m plate-girder spans was made in aluminium instead of steel as an experiment (1938). All these early structures were in naturally aged 2xxx-series alloy (duralumin) that would be thought unsuitable by modern ideas. The Smithfield Street deck lasted for 40 years, despite the choice of alloy, after which it was replaced by another aluminium one. This second deck is built of 6xxx-series extrusions, using welded construction.

Immediately after the war, steel was scarce and an available alternative was aluminium, which was specified for various structures, such as factory roofs and cranes, despite the extra cost. Much of this market soon disappeared, although aluminium continued to appear in roofs of large span [6]. Today it still appears in some such roofs, typically employing proprietary 'space-frame' forms of construction.

In bridges, the greatest consumption of aluminium has been in military bridges, which have to be erected and launched in record time. Welded aluminium kits designed for this use are a worthy successor to the famous Bailey, and have consumed thousands of tonnes. Aluminium also appears in civilian bridges in remote locations, where they are left unpainted. There is a theoretical case for aluminium in bridges of very large span, where self-weight is a major factor, but such a development has yet to appear.

A more recent heavy structural development is in the offshore field. Here aluminium is gaining acceptance as a valid material for modules on fixed platforms, where the cost of installation critically depends on weight. Typical examples are helidecks and accommodation modules. The latter may be described as all-aluminium five-storey hotels, that are floated out and lifted into place in one piece, complete with cinema.

In the late 1940s, some designers followed American earlier practice by specifying 2xxx-series alloy for large structures. In a few cases this was a disastrous choice. For example, a bascule bridge in such material at the docks in Sunderland, UK, failed within a few years, due to corrosion in the severely polluted marine/industrial atmosphere. Another example was British crane jibs designed in the stronger form of 2xxx (superdural), using stress levels that were much too high. These failed by fatigue after a very short life. It was soon realised that the best alloy for civil engineering structures is usually the stronger type of 6xxx-series material, although the weldable kind of 7xxx is sometimes preferred. A notable example of 7xxx usage is in military bridges.

1.5.4 Architecture

An early installation of aluminium window-frames was on the University Library at Cambridge in 1936. Since then aluminium has become widely used in windows and other secondary components, including curtain-walling, patent-glazing and conservatories. Typical practice is to design a 'suite' of special extrusions, which enable assembly to be made by means of secret-fix mechanical joints instead of welding. The standard alloy is the weaker 6xxx type with its good surface finish, stiffness rather than strength being the main design factor. In recent years, aluminium window-frames have begun to lose out to plastic in the domestic market.

Aluminium sheeting competes with powder-coated steel for the cladding of factory buildings, typically using a roll-formed trapezoidal profile in 3xxx-series alloy sheet. A different kind of application is in 'fully-supported' roofing, where very high purity soft aluminium sheet is a valid alternative to lead, zinc or copper.

1.5.5 Land transport

Since 1945 aluminium has steadily replaced steel in railway carriages. Earlier construction consisted of conventional framing with applied sheet metal panels. This is now giving way to designs that employ wide hollow extrusions to produce an efficient double-skin all-welded form of construction. The stress levels have to be kept low, because of fatigue, making it economic to use the weaker sort of 6xxx alloy with its good extrudability. Aluminium has also been successfully employed in the construction of large ore-wagons. In North America some 90% of such wagons are made of aluminium.

In road transport, aluminium has a long established market for the bodies of lorries and vans, which are mounted on a conventional steel chassis. Ingenious suites of extrusions are available, which enable the small fabricator to build individual bodies to suit different customers. Such 'Meccano sets' have not changed much since they were first introduced around 1950, an important item being the interlocking 'plank' sections used for the floor. Standard material is the stronger 6xxx type, and assembly is by bolting and riveting with a limited amount of welding. A different technique applies with tipper bodies where all-welded construction is the norm, using 5xxx-series plate. Efforts to build the chassis in aluminium go back to 1950, but with limited success. More recently we have seen some more advanced goods chassis prototypes, using welded and also bonded construction.

These transport developments are likely to be dwarfed by another that is on the verge of taking off – the motor car. This potentially vast market has been slow to develop, despite the obvious advantage of a light, strong, non-rusting aluminium body. Hitherto the metal has had a very limited role in cars, just at the expensive end of the market and in some sports cars. But advanced designs are now appearing, assembled with a combination of welding and adhesive bonding, which enable body-shells to be produced that are comparable to steel in terms of overall stiffness and crash damage absorption. It is claimed that such bodies are no more expensive, because the higher metal cost of the aluminium is offset by cheaper finishing, the integrity of the paint system being much less critical than it is with steel (which rusts). With such designs aluminium is poised to move into the high-production end of the market, and already one such model (Audi) is being sold. It is conceivable that the people's car of the future will have an unpainted aluminium body, on which the presence of scratches and dents can be ignored.

1.5.6 Marine usage

A major marine application has been for the superstructures of liners, where reduced weight up-top aids stability. Well-known examples of such liners were the *QE2* and the *France*, the latter complete with aluminium squash court. For the same reason, aluminium is sometimes chosen for the superstructures of warships. Since ductility is as important as strength, the preferred material is the stronger form of 5xxx-series plate. The extruded stiffeners used to be in the same material, but current practice tends toward the more extrudable 6xxx type.

Recent years have seen the introduction of high speed multi-hull ferries such as the Australian designed catamaran *Sea Cat* on the Channel run.

These tend to be in all-aluminium construction, again using a combination of 5xxx-series plating and 6xxx extrusions.

1.6. SOURCES OF INFORMATION

The following are some useful books published in recent years on the structural use of aluminium.

R. Narayanan (ed.) (1987)	*Aluminium Structures: Advances, Design and Construction*. Elsevier Applied Science, Amsterdam.
H. Spencer (1989)	*Aluminium Extrusions, A Technical Design Guide*. The Shapemakers (UK Aluminium Extruders Association), Broadway House, Calthorpe Road, Birmingham, UK.
J. Lane (1992)	*Aluminium in Building*. Ashgate Publishing, Aldershot, UK.
P.S. Bulson (ed.) (1992)	*Aluminium Structural Analysis, Recent European Advances*. Elsevier, Amsterdam.
M.L. Sharp (1992)	*Behaviour and Design of Aluminum Structures*. McGraw-Hill, New York.
J.W. Bull (1994)	*The Practical Design of Structural Elements in Aluminium*. Ashgate Publishing, Aldershot, UK.
F.M. Mazzolani (1995)	*Aluminium Alloy Structures* (second edition). E & FN Spon, London.
J.R. Kissell and R.L. Ferry (1995)	*Aluminum Structures, A Guide to Their Specification and Design*. Wiley, New York.

By far the most comprehensive collection of training material on aluminium is a CD-ROM entitled *TALAT (Training in Aluminium Technologies)*. This comprises 129 lectures contributed by experts from 11 European countries. It covers materials technology, design, surface technology and all aspects of fabrication. It is available from member associations of the European Aluminium Association (EAA) as follows.

Britain	Aluminium Federation Ltd (Birmingham)
Germany	Aluminium Zentrale eV (Dusseldorf)
Netherlands	Stichting Aluminium Centrum (Woerden)
Scandinavia	Skanaluminium (Oslo)
Other countries	EAA (Brussels)

Erection of 46 m span portal-frames for aircraft hangars at Heathrow Airport, London, 1950. Battened channel construction, riveted. Built by S.M.D. Engineers of Slough.

Aluminium roof structure on 80 m span warehouse for Compagnie Maritime Belgique, Antwerp, 1957. Back-to-back bulb-angle construction, using large cold-driven rivets.

Aluminium roof structure for exhibition hall in Rio de Janeiro, 1978, using Professor Marsh's 'M-dec' system.

Close-up of the M-dec system, which employs a suite of special extrusions.

Aluminium structure, in the form of a 28 m cube, used for enclosing an auditorium in Medellin, Columbia, 1990.

Aluminium structure for the 51 m diameter dome on the Selangor State mosque, at Shah'Alam, Malaysia, 1987. Built by BACO Contracts, using their 'Triodetic' system.

Space-frame joint in BACO's 'Triodetic' system, using a special node extrusion with serrated grooves.

Aluminium glazing system for commercial greenhouses, built from a suite of extrusions including a special valley gutter section. (Courtesy of Cambridge Glass Co.)

The 'BR90' bridging system, which entered service with the British Army in 1997. The trussed version shown here is carrying the Challenger battle tank on a 56 m span. Built by Vickers Defence Systems at Wolverhampton, using 7xxx-series alloy.

One of the aluminium accommodation modules for the Saga oil-field in the North Sea, before float-out. Welded up from 0.5 m wide extrusions by Leirvik Sveis on Stord Island, Norway, 1990.

Aluminium helideck installed on a production platform in the North Sea. Constructed of wide hollow extrusions.

500-kV guyed aluminium transmission tower in Canada. (Courtesy of Professor Marsh, Montreal.)

Nat West Media Centre, Lords Cricket Ground, London, 1998. Welded construction using 5xxx-series plate. The main contractor was Pendennis Shipyard Ltd, Falmouth, Cornwall. (Copyright Ove Arup and partners; photograph: Peter Mackinven)

CHAPTER 2

Manufacture

In this chapter we describe the manufacture of wrought aluminium products, such as plate, sheet, sections and tube, known in the trade as *semi-fabricated products* (or 'semi-fabs'). The two main stages are production of pure aluminium ingot, and conversion of this into wrought material.

2.1 PRODUCTION OF ALUMINIUM METAL

2.1.1 Primary production

The production of aluminium ingot comprises three steps: (1) mining the bauxite ore; (2) extracting alumina therefrom; and (3) smelting.

Bauxite is plentiful in many countries and is extracted by opencast mining. Alumina (Al_2O_3), a white powder, is obtained from bauxite by the Bayer process which requires a large supply of coal and caustic soda. Roughly 2 kg of bauxite, 2 kg of coal and 0.5 kg of caustic soda are needed to produce 1 kg of alumina.

In the smelting operation, metallic aluminium is extracted from alumina by electolysis, using the Hall–Héroult process. It takes place in a cell (or 'pot') comprising a bath of molten cryolite (the electrolyte) and carbon electrodes. A typical pot is around 4 or 5 m long. The cathode covers the floor of the bath, while the anodes are in the form of massive carbon blocks which are gradually lowered into the cryolite as they burn away.

Alumina is fed in at the top of the molten pool, where it dissolves, and molten aluminium is drawn off from the bottom. The metal emerges at a temperature of around 900°C and at a high purity (99.5–99.8%). Copious fumes are produced. The electric supply is direct current at a typical potential of 5 V per pot, the current requirement being high, around 100–150 kA. Continual replacement of the carbon anodes is also a significant factor. The production of one kilo of aluminium consumes roughly 2 kg of alumina, 0.5 kg of carbon and 15 kWh of electrical energy. The cryolite (Na_3AlF_6) is largely unconsumed.

The main requirements for the production of aluminium are thus bauxite, coal and cheap electricity. These are never all found in one place. The location usually preferred for a smelter is near a dedicated hydroelectric power plant, which may be thousands of kilometres from the alumina source and from markets for the ingot, as, for example, with the Kitemat smelter in British Columbia, Canada. Sometimes a smelter is designed for strategic reasons to run on coal-fired electricity. The new 50 000 tonne/yr smelter at Richards Bay in Natal, South Africa, which relies on coal-fired electricity, has a total of nearly 600 pots contained in four 'potrooms' each 900 m long × 30 m wide.

2.1.2 Secondary metal

Not all the metal going into aluminium products comes from ingot, an important ingredient being 'secondary metal', i.e. scrap. This is partly supplied by scrap merchants, and partly comes from process scrap generated in the rolling mill or extrusion plant. The composition of such scrap is important, the best scrap being pure aluminium or a low alloy. Melted-down airframe material is less convenient, as it contains a relatively large amount of copper or zinc, making it less suitable for making non-aeronautical alloys.

2.2 FLAT PRODUCTS

2.2.1 Rolling mill practice

Aluminium plate and sheet are manufactured in conventional rolling mills, the main difference from steel being the lower temperatures involved.

Alloyed metal is produced by melting a mixture of ingot and scrap, to which are added metered quantities of 'hardeners' (small aluminium ingots containing a high concentration of alloying ingredient). *Rolling slabs* are then produced by vertical continuous casting in a long length, from which individual slabs are cut. Each slab is skimmed on both faces and slowly heated in a furnace, from which it enters the hot-line. This might typically comprise a hot mill followed by a three-stand tandem set-up. The material emerging therefrom, called 'hot-mill strip', can be shipped directly as plate or reduced further in a cold mill to produce sheet. The immediate output from the cold mill is coiled strip, which is decoiled and cut into sheets. Plate and sheet widths thus produced are typically about 1.5 m, although greater sizes are possible.

2.2.2 Plate

The term 'plate' refers to hot-rolled flat material, typically with a thickness $\geqslant 6$ mm, although it may be thinner. When in a non-heat-treatable alloy, plate is often supplied in the rather vague 'as-hot-rolled' or F condition, for which only typical properties can be quoted. Alternatively it may be annealed after rolling, bringing it to the O condition, in which case the strength is lower but clearly specified. It is also possible for non-heat-treatable plate to be rolled to a temper (i.e. specific hardness), although this can be difficult to control on the hot-line.

Heat-treatable plate (2xxx, 6xxx or 7xxx-series alloy) is usually supplied in the fully-heat-treated T6-condition, which involves a quenching process followed by artificial ageing. Alternatively it can be produced in the more ductile T4-condition, when it is allowed to age naturally at room temperature.

2.2.3 Sheet

By 'sheet' we generally refer to flat material up to 6 mm thick (although in the USA the term 'shate' is sometimes used for material on the border-line between sheet and plate). It is produced by cold reduction, which usually involves several passes with interpass annealing. Readily available thicknesses are 0.5, 0.6, 0.8, 1.0, 1.2, 1.5, 2.0, 2.5, 3.0, 4.0, 5.0 and 6.0 mm.

Most sheet is produced in non-heat-treatable material (1xxx, 3xxx, 5xxx), and is supplied in a stated *temper* (such as 'quarter-hard', 'half-hard') with specified mechanical properties. These are achieved by a controlled sequence of rolling reductions, possibly with interpass annealing. The sheet may be *temper rolled*, in which case it is given a precise reduction after the last anneal. Or it may be *temper annealed*, when the final stage is a suitably adjusted anneal after the last rolling pass.

For more highly stressed applications, sheet is supplied in heat-treatable alloy (2xxx, 6xxx or 7xxx series), usually in the fully-heat-treated T6-condition (quenched and artificially aged).

2.2.4 Tolerance on thickness

The thickness tolerance δt for flat products allows for variation across the width, as well as for inaccuracy in the main thickness. Its value depends on the thickness t and width w, and may be estimated approximately from the following expressions:

$$\text{Plate} \qquad \delta t = \pm\,(0.0000\,14\,wt + 0.3)\ \text{mm} \tag{2.1}$$

$$\text{Sheet} \qquad \delta t = \pm\,(0.11\sqrt{t} + 0.00004\,w - 0.06)\ \text{mm} \tag{2.2}$$

in which t and w are in mm. These agree reasonably with the BSEN.485

requirements for $t = 1–30$ mm and $w > 1000$ mm. Outside these ranges, they overestimate the tolerance.

2.2.5 Special forms of flat product

(a) Clad sheet

Sheet material in 2xxx-series alloy, and also in the stronger type of 7xxx, can be produced in a form having improved corrosion resistance by cladding it with a more durable layer on each surface. Such a product is achieved by inserting plates of the cladding material top and bottom of the slab as it enters the hot-mill. As rolling proceeds, these weld on and are steadily reduced, along with the core, the proportion of the total thickness being about 5% per face. For 2xxx-series sheet, the cladding is in pure aluminium, while for the 7xxx an Al–1% Zn alloy is preferred.

(b) Treadplate

Aluminium treadplate is available with an anti-slip pattern rolled into one surface, as in steel. It is normally supplied in the stronger type of 6xxx-series alloy in the T6-condition.

(c) Profiled sheeting

This product, which can be used for many sorts of cladding, is made by roll-forming. An important use is for the cladding of buildings, where a trapezoidal profile is usually specified, formed from hard-rolled sheet in 3xxx-series alloy. Some such profiles are of ingenious 'secret-fix' design.

(d) Embossed sheet

This is sheet having a degree of roughening (of random pattern) rolled into one surface. It can be used in order to reduce glare. It is also claimed to improve the stiffness slightly. Such sheet is sometimes employed in the manufacture of profiled sheeting.

(e) Cold-rolled sections

As in steel, it is possible to produce small sections by roll-forming from strip. However, the technique is not generally favoured for aluminium, because extrusion has more advantages. But it comes into its own for very thin sections, of thickness less than the minimum extrudable, say, 1 mm and under.

2.3 EXTRUDED SECTIONS

2.3.1 Extrusion process

Although available for some other non-ferrous metals, such as brass and bronze, it is with aluminium that the extrusion process has become a major manufacturing method, far more so than with any other metal [7]. This is partly due to the relatively low temperature at which the metal will extrude (roughly 500°C).

The process enables aluminium sections to be produced from 10 to 800 mm wide, with an unlimited range of possible shapes. The tool cost for producing a new section is a minute fraction of that for a new section in steel, as it merely involves cutting a new die. Also, the down-time at the press for a die change is negligible compared to the time lost for a roll change in the steel mill. In aluminium, it is therefore common practice to design a special dedicated section to suit the job in hand, or a 'suite' of such sections, and the quantities do not have to be astronomic to make this worthwhile. An important feature of aluminium extrusion is the ability to produce sections that are very thin relative to their overall size.

There are various versions of the extrusion process. Aluminium sections are normally produced by *direct extrusion*. As for flat products, the starting point is molten metal with a composition carefully controlled by the addition of hardeners. Long cylindrical 'logs' are then produced by continuous casting, often at the smelting plant. These are cut into shorter lengths by the extruder to produce the actual extrusion billets. Previously these were always skimmed in a lathe, to remove surface roughness and impurities, but modern practice is to dispense with this in the case of 6xxx-series billets, because of improvements in log casting. Each billet is preheated in an induction furnace and then inserted in the heated container of the extrusion press (Figure 2.1). The hydraulic ram at the back of the billet is then actuated, causing the metal at the front end to extrude through the die and travel down the run-out table. The aperture in the die defines the shape of the emerging section. The process is continued until some 85–90% of the billet has been used. The as-extruded length may reach 40 m.

The *extrusion ratio* is the ratio of the section area of the billet to that of the section extruded. For 6xxx-series alloys, it ideally lies in the range 30–50. Too low a value (say, 7 or less) will cause a drop in properties; while too high a value (say, 80 or more) means an excessive ram pressure, with the possibility of die distortion and breakage.

Extrusion presses vary greatly in size, with the container bore (billet diameter) ranging from about 100 to 700 mm. The required pressure from the ram depends on the alloy and the extrusion ratio. Presses are rated according to their available ram force, which typically lies between 10 and 120 MN (1000 and 12 000 tonnes). Although presses are mostly located in

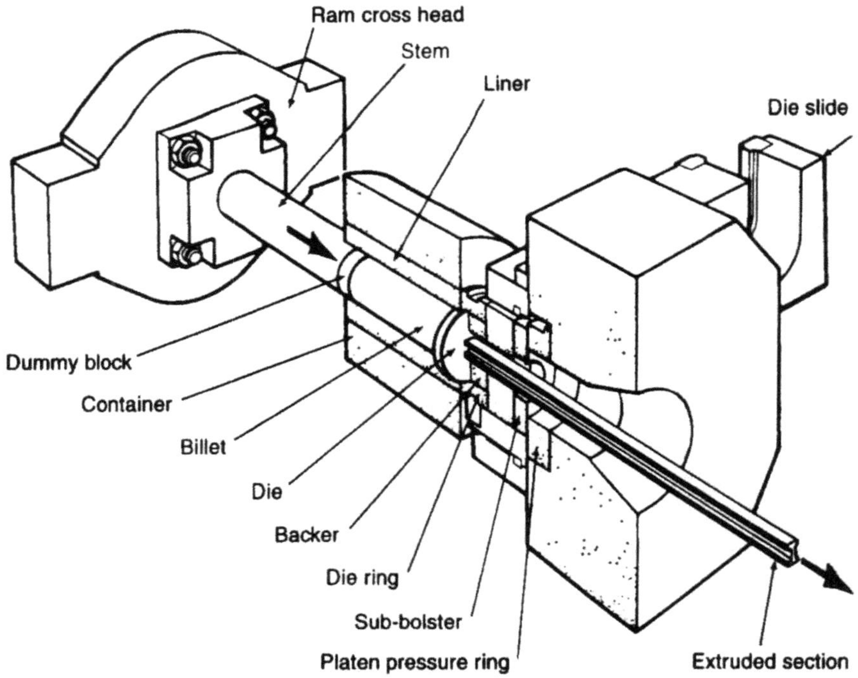

Figure 2.1 Extrusion process (direct extrusion).

big extrusion plants, it is not uncommon to find a small press in the factory of a specialist fabricator, such as a metal window firm.

2.3.2 Heat-treatment of extrusions

Most extrusions are produced in heat-treatable material, and to bring them up to strength they have to undergo solution treatment (quenching) followed by ageing.

The easiest form of solution treatment is simply to spray the section with water as it emerges from the press, and this is the usual procedure for thinner sections in 6xxx-series alloy. With some 6xxx material, a useful degree of hardening is even achieved with the spray switched off ('air-quenching'), thereby reducing distortion.

Quenching at the press is less effective with thick 6xxx material, and with the 2xxx and 7xxx-series alloys it is no good at all, because these require precise control of the solution treatment temperature. For such material, it is necessary, after cutting into lengths, to reheat and quench in a tank. This can be done vertically or horizontally. The former causes less distortion, but imposes a tighter limitation on length.

For most extruded material, the second stage of heat treatment (the ageing) consists of holding it in a furnace for some hours at an elevated temperature somewhere in the range 150–180°C. This is known as *artificial ageing* or *precipitation treatment*, and takes the metal up to its full strength T6 condition (or T5 if air-quenched). It is performed after correction. Sometimes the quenched material is left to age naturally at room temperature (*natural ageing*), bringing it to the more ductile but weaker T4 condition. This would be preferred for material that has later to go through a forming operation.

2.3.3 Correction

Extrusions tend to distort as they come off the press, and the quenching operation makes this worse. There are basically two forms of distortion that have to be corrected: (a) overall bow along the length; and (b) distortion of the cross-section.

Overall bow is got rid of by stretching, typically up to a strain of 1 or 2%. For spray or air-quenched material, and also for non-heat-treated, stretching can take place on the full length of the section as extruded, before it is cut into shorter lengths. For other materials, it has to be done length by length after quenching. For heat-treated extrusions, the stretch has little effect on the final material properties. But for non-heat-treatable extrusions, it has the effect of significantly lowering the compressive proof stress (the Bauschinger effect), a fact that is apt to be ignored by designers.

For thick compact sections, distortion of the cross-section is no problem, and the only correction they need is the stretch. But for thin slender ones this form of distortion can be serious and further correction is needed. Various techniques are available, expecially roller correction, and these are tailored to suit the profile concerned. These techniques tend to be labour intensive and, for this reason, slender sections cost more per kilogram. Sometimes the likelihood of serious distortion in a very slender profile will make a proposed section impracticable, even though it can be extruded. In such cases, a possible answer may be to reduce the distortion by specifying the air-quenched T5 condition (instead of T6), provided the lower properties are acceptable.

2.3.4 Dies

Extrusion dies, which are made by the thousand, are in a special hard heat-resisting steel, the aperture being machined by spark erosion. The tooling cost for a straightforward structural die (non-hollow), 150 mm wide and without complications, might be roughly one-third of the cost of a tonne of the metal supplied from it.

Skill is needed to make a die so that the section produced comes out more or less straight and level down the run-out table. Referring to Figure 2.2, it is the aperture dimension at the entry side that controls the thickness. The thick parts of a section tend to extrude faster than the thin parts,

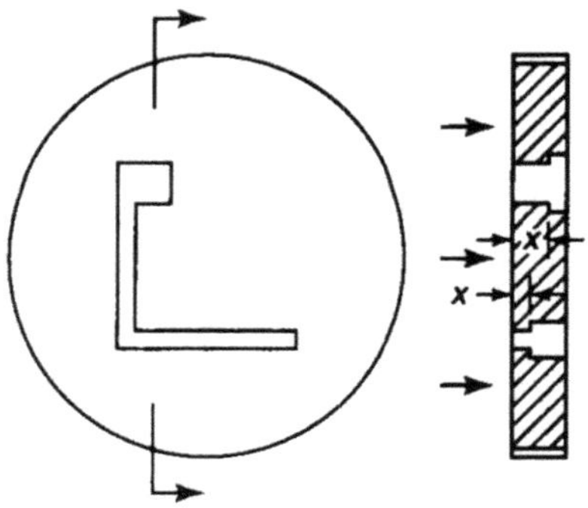

Figure 2.2 Typical extrusion die.

so that a section such as the one shown would come out in a curve if nothing were done. To counter this effect, the die designer retards the flow of metal in the thick regions by increasing the depth of 'land' (dimension x). Even so, a section never comes off the press completely straight and subsequent stretching is always needed.

The performance of a die can be improved if any re-entrant corners in the aperture (outside corners on the section) are slightly radiused, even with a radius of only 0.3 mm. It is bad practice to call for absolutely sharp corners unless these are essential. They increase the risk of die failure and reduce the permitted extrusion speed.

The pressure acting on the face of a die during extrusion is very high, possibly approaching 700 N/mm^2. With sections such as those shown in Figure 2.3, there is the possibility that the die will break along line Y due to the pressure acting on region X. This tendency depends on the *aspect ratio* (α) of the region X, defined by:

$$\text{Plain slot} \quad \alpha = \frac{d}{c} \qquad (2.3a)$$

$$\text{Re-entrant areas generally} \quad \alpha = \frac{A}{c^2} \qquad (2.3b)$$

where c, d are as defined in the figure and A is the area of region X. Under

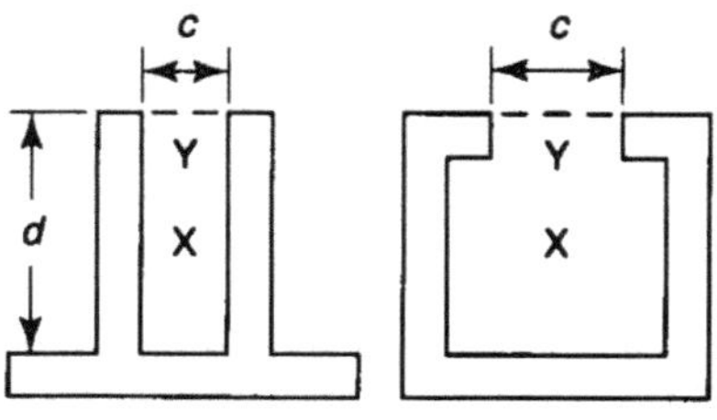

Figure 2.3 Re-entrants in extruded sections.

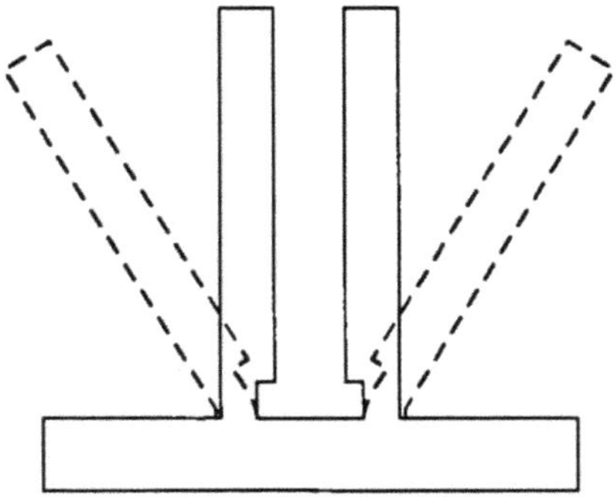

Figure 2.4 Roller modification of profile with deep re-entrant.

the most favourable conditions, i.e. for a section in 6063 alloy (or pure aluminium) having rounded corners at the tips, it may be assumed that such extrusions are viable provided the aspect ratio is less than about 3.0. This limiting value tends to decrease slightly with the stronger types of 6xxx material, or if the tip corners are not rounded. It decreases further for the weaker (weldable) forms of 7xxx-series alloy, and much more so for 2xxx, 5xxx and the stronger 7xxx alloys.

When α exceeds the limiting value, there are two possible courses. The first is to extrude the section in an opened-out shape and then roll-form it back to the desired profile during correction (Figure 2.4). The alternative is to extrude it as a quasi-hollow, or 'semi-hollow', using a bridge die (see below).

2.3.5 Hollow sections

Hollow sections are normally extruded using a two-piece *bridge die* (Figure 2.5). The outside of the section is defined by the aperture in the front part A of the die, and the inside by the mandrel nose on the back part

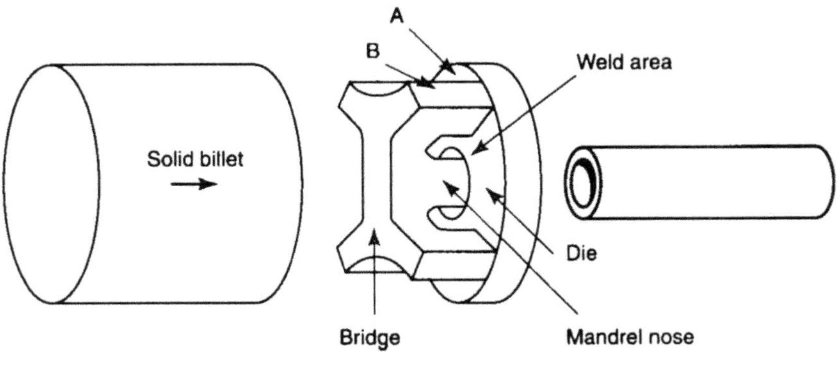

Figure 2.5 Production of hollow section using a bridge die.

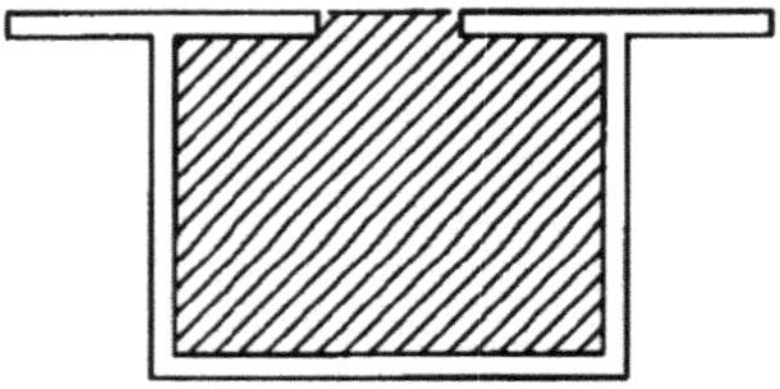

Figure 2.6 'Semi-hollow' profile.

B. The extrusion speed is a bit less than for a non-hollow profile, leading to a slightly higher cost per kilogram. Also, the cost of the die is likely to be twice as much as for a non-hollow profile, so that a reasonable size of order is needed to make a new section worthwhile.

Bridge dies are also employed for extrusion of 'semi-hollow' profiles, such as that shown in Figure 2.6. For these, the nose on part B of the die defines the area shown shaded in the figure.

During the extrusion of a hollow section, the plastic metal has to flow around the support feet of part B and then reunite before emerging. The section thus produced therefore contains local zones at which welding has occurred during extrusion, called 'seam welds'. These are usually of no consequence and many users do not even know they are there. But a designer should be aware of them, since they produce potential lines of weakness down the length of a section which very occasionally cause trouble, especially if the extrusion speed is too high. With the section shown in Figure 2.7, to which web-plates are to be welded by the fabricator, there is a risk that transverse shrinkage at these welds might tear the section apart at one of the seam welds, if the latter are located as shown. When ordering such sections, it is prudent to tell the extruder where any fabrication welds will be made, so that the seam welds can be located well away from them.

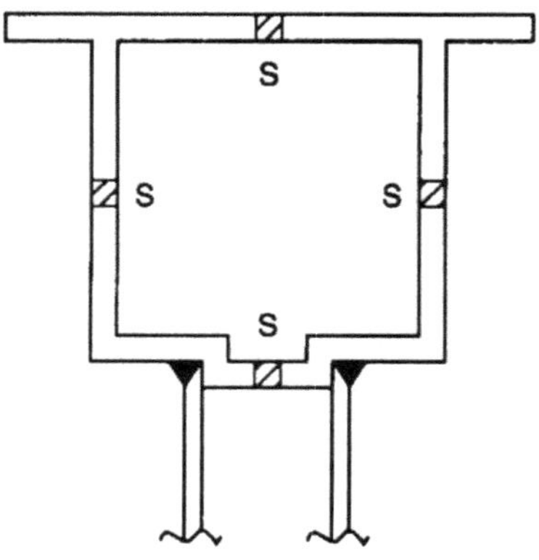

Figure 2.7 Seam welds (S) in a hollow profile.

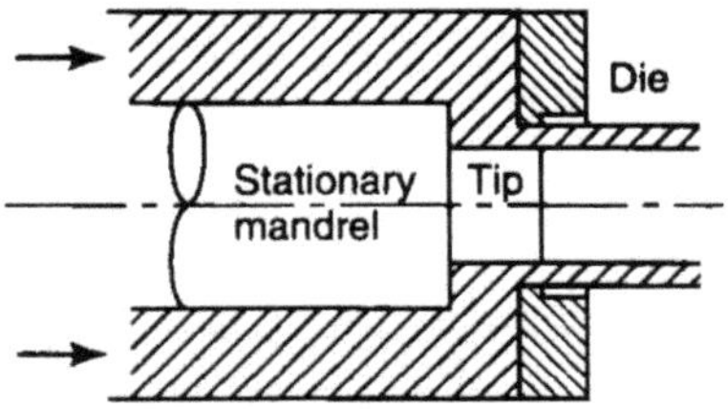

Figure 2.8 Production of hollow section by extrusion over a mandrel.

An alternative technique for producing hollow shapes, which eliminates seam welds, is 'extrusion over a mandrel' (Figure 2.8). A suitable press is needed for this (a 'piercing press') in which the main ram is bored out to accommodate a separately controlled inner one. A simple die is used to define the outside of the section, while the internal profile is generated by a tipped mandrel carried on the inner ram and passing through a hole in the billet. The method is slow to set up and more costly. Also there is a tendency for the long mandrel to deflect sideways, causing eccentricity of the interior of the section relative to its exterior. The technique is used when the very slight risk from the presence of seam welds is unacceptable.

2.3.6 Extrudability of different alloys

The extrusion alloys *par excellence* are the 6xxx series. They extrude nearly as readily as pure aluminium and are highly suitable for the production of slender profiles, especially the weaker type such as 6063. They are also ideal for the extrusion of hollow sections through a bridge die. They have the further advantage that reliable properties can be achieved by quenching at the press (spray quenching). Extrusion speeds of 25 m/min or more are quite normal with 6063 extrusions.

The weaker (i.e. weldable) type of 7xxx-series material also extrudes well, and is again suitable for hollow shapes. But such material extrudes more slowly, and the same degree of section slenderness cannot be achieved as with 6xxx material. Also the solution treatment has to be done as a separate operation, and not by spray quenching at the die.

The 2xxx-series alloys, and also the stronger type of 7xxx alloys, are altogether less convenient for extrusion. They extrude slowly and are no good for the production of hollows using bridge dies.

Extruded sections in non-heat-treatable aluminium (1xxx, 5xxx series) are of limited use for structural profiles, because they have to be supplied in the relatively soft as-extruded F condition; there is no way of hardening them by cold work, extrusion being a hot process. Therefore they are only called for when high ductility is needed. Pure aluminium extrudes very readily. The 5xxx-series alloys are generally poor for extrusion, especially the stronger alloys in the group (such as 5083). These come out slowly from the die, and cannot be used with bridge dies.

2.3.7 Size and thickness limits

Precise data on the limiting geometry for extruded profiles is impossible in a book such as this, and we can only give general guidance. It is obviously essential to confer with the extruder if there is any chance that a proposed section may cause problems.

(a) Length

For material that is unsuitable for spray quenching, such as 2xxx or 7xxx-series material, the maximum length is governed by the heat-treatment set-up. Typical rough figures might be 6 m for vertical quenching, and 10 m for horizontal. This restriction does not apply with 6xxx-series material quenched at the press, for which the probable limitation is transport. With a very thick section, the available length may be restricted by the amount of metal in the billet.

(b) Width

There is no difficulty in obtaining extrusions up to 300 mm wide, for which there are many presses with a large enough container. Twice this width is possible from the largest presses, 600 mm or more, but these exist in only a few countries.

(c) Thickness

A designer will often want to make a section as thin as possible, for economy. The minimum thickness depends partly on what can be extruded, depending on the alloy, and partly on the control of distortion. It is generally a function of the overall size of the section. As a very rough guide, the minimum practicable thickness for a non-hollow in the weaker type of 6xxx-series alloy (such as 6063) is the greater of (a) 1.2 mm; and (b) 1% of the section size.

It is emphasized that these are approximate values, and that the minimum possible thickness for a given profile may sometimes turn out to be thicker and sometimes thinner. Value (b), which refers to general thickness rather than small ribs, etc., is more likely to be achieved when the T5 condition is specified (air quenched). For hollows, the minimum thickness will probably be some 25% higher than for non-hollows.

For the stronger type of 6xxx material, the minimum available thickness tends to be slightly greater. For the weaker (weldable) type of 7xxx material, it is significantly greater; while for 2xxx material, and also for the stronger versions of 7xxx and 5xxx material, it is considerably so.

2.3.8 Tolerances

When one sees a batch of gleaming aluminium extrusions, one is tempted to think of them as a precision product, although in fact they are subject to tolerances in the same way as steel. Such tolerances cover width, thickness, section shape, straightness and twist. When a particular dimension is critical, it is usually possible to negotiate with the extruder a tighter figure for the tolerance concerned. An example would be when two sections have to mate together. It is beyond our scope to cover tolerances in detail, and the following is a rough guide.

The normal tolerance on section width w may be estimated from the following approximate expression:

$$\text{Width tolerance} \approx \pm\left(0.3+\frac{w}{200}\right)\text{mm} \qquad (2.4)$$

where the width w is in mm. This agrees reasonably with the British Standard BSEN.755 requirements when $w > 100$ mm, erring on the high side for low values of w. An equivalent expression for the tolerance on thickness t is:

$$\text{Thickness tolerance} = \pm\left(0.2+\frac{w}{600}+\frac{t}{150}\right)\text{mm} \qquad (2.5)$$

where w and t are in mm. This refers to the thickness of the main elements of the section, and does not apply to small fins etc. Again it tends to be pessimistic at low w.

Expression (2.5) can give disturbing results, if taken at face value. For example, if $w = 300$ and $t = 3$ mm, the estimated tolerance would be ± 0.7 mm. Applied over the whole section this could lead to an area some 20–25% above nominal, with a corresponding increase in the weight. If the section was being bought at a fixed rate per kilogram, the effect on the economy of the design would be disastrous. In fact, the tolerance value given by equation (2.5) mainly covers local variations in thickness around the section. The reason it increases so much with w is to allow for flexing of the die in a wide thin section. Such flexing affects the thickness at the middle of a wide thin element. At the edges the variation will be much less, and can be estimated by ignoring the second term in expression (2.5). Therefore the possible variation in cost per metre will be much less than might at first be supposed. But it is still significant.

An important factor governing variation in thickness is die wear. On a long run, the die has to be taken out of service when it gets too worn and replaced by a fresh one. A tighter tolerance can be achieved by limiting the interval between such replacements.

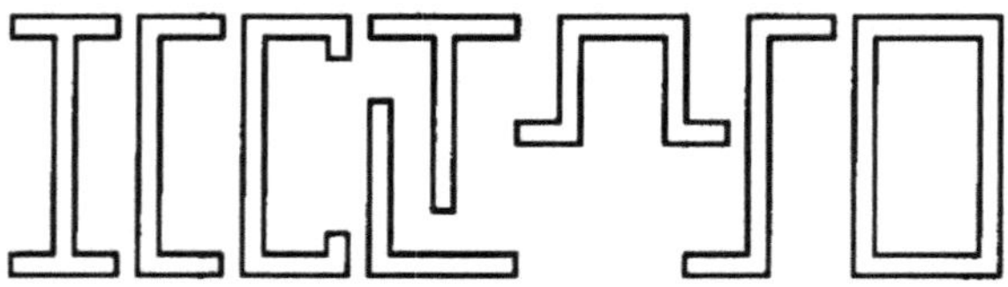

Figure 2.9 Conventional profiles.

2.3.9 Design possibilities with extrusions

A designer has the option of either using an existing section when a suitable die already exists, or getting a new die cut to produce the optimum section for the job. Extruders carry a large range of existing dies for sections of conventional shape, like those shown in Figure 2.9, and many such sections are held by stockists.

When designing a new section it is sensible to combine functional features with structural strength, thus designing out some of the fabrication.

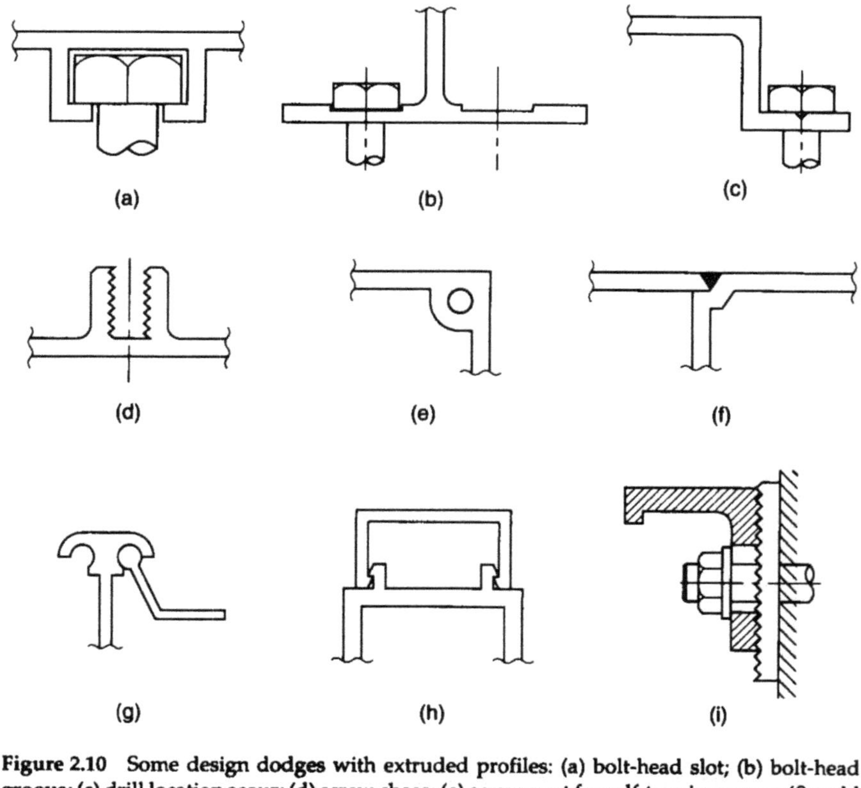

Figure 2.10 Some design dodges with extruded profiles: (a) bolt-head slot; (b) bolt-head groove; (c) drill location scour; (d) screw chase; (e) screw-port for self-tapping screw; (f) weld prep; (g) hinge sections; (h) snap-on cap; (i) positive-fix adjustable mounting (with slotted hole).

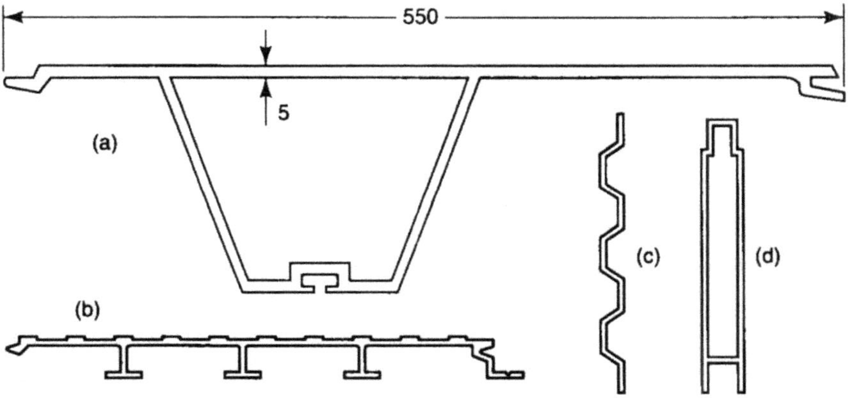

Figure 2.11 Some typical plank sections: (a) flooring for offshore accommodation module; (b, c, d) truck body sections.

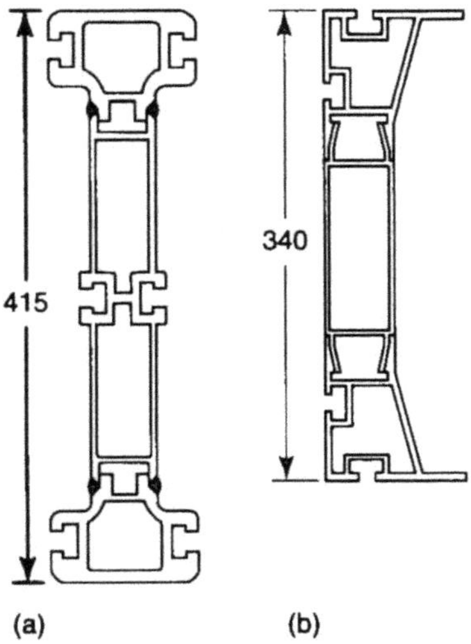

Figure 2.12 Two designs of chassis member using special extrusions: (a) welded; (b) bonded. Note use of bolt-head slots for the connection of cross-members, thus improving the fatigue rating.

Figure 2.10 shows some of the well-known tricks of the trade. Plank sections are a widely used class of profile (Figure 2.11) with various possible features such as anti-slip surface, concealed fixing and positive interlock between adjacent planks. Figure 2.12 shows two designs by British engineer Ron Cobden for truck chassis members, one welded and the other bonded. An important feature in each is the use of bolt-slots to receive the heads of the bolts connecting the cross-members, thus improving the fatigue performance. Free advice on the design of extruded profiles may be had from The Shapemakers (page 16).

2.4 TUBES

By *tubes* we mean hollow sections of uniform thickness, usually round, but sometimes shaped (square, hexagonal, etc.). There are essentially three ways of making such tubes, extrusion, drawing or welding.

2.4.1 Extruded tube

This product is simply a hollow extrusion produced using a bridge-die (see Section 2.3.5). It is typically supplied in 6xxx-series alloy or the weaker kind of 7xxx alloy. It has the disadvantage of not being truly seamless, because of the seam-welds, and this bars its use for some applications.

2.4.2 Drawn tube

Often referred to as *seamless tube*, this is a high quality product costing more than extruded tube per kilogram. It starts off as a relatively thick hollow 'bloom', produced by extrusion over a mandrel. This is then reduced by a succession of passes on a draw-bench, with inter-pass annealing when necessary. The reduction operation consists of gripping the tube by its 'tagged' end and drawing it through a die. The outside diameter (OD) is defined by the die, and the inner diameter (ID) by a plug carried on a long rod passing back through the tube. The operation is relatively labour intensive, as it has to be done length by length. For some applications, however, it is possible to draw coiled tube in long lengths, using a floating plug.

Drawn tube can be readily produced in thin gauges, below what can be extruded. Its main advantage is being seamless. It also enables the OD and ID to be held to close limits, and the surface finish is good. On the debit side, it tends to exhibit eccentricity of the bore, due to the way in which the blooms are made. This results in a thickness variation around the periphery, the typical tolerance on thickness at any one point being 10% of nominal. One way to overcome this is to use a bloom extruded through a bridge-die, but then the final product is not truly seamless.

Because the process involves cold work, drawn tube is suitable for non-heat-treatable as well as heat-treatable material. With the former, it is supplied to a specified temper (degree of hardness).

Shaped drawn tube (i.e. non-circular) is made by first reducing a bloom in the round to the necessary diameter and thickness, and then drawing it through a shaping die to form the final profile.

2.4.3 Welded tube

This is the cheapest way of making thin tube. Cold-reduced strip is roll-formed to a circular shape and then fed past a welding head to produce the final product. Such tube is produced in long lengths from non-heat-treatable coiled strip (3xxx or 5xxx-series alloy). A major market is for irrigation pipe.

CHAPTER 3

Fabrication

3.1 PREPARATION OF MATERIAL

3.1.1 Storage

Aluminium materials waiting to be fabricated are obviously best stored in the dry. But if they do get wet, at least they will not rust. One possible problem during storage is *water staining*, which appears as ugly blotches varying from white to dark grey [8]. This occurs when water is trapped between flat aluminium surfaces, and sometimes happens as a result of condensation when metal is brought in from the cold. Water stains have no effect on structural performance, but must be avoided when good appearance is needed.

3.1.2 Cutting

Most of the techniques used for cutting steel are suitable for aluminium, including shearing, sawing, notching, nibbling and routing. Sawing, in particular, is quicker and easier than for steel, provided the appropriate tooth size and cutting speed are used. Circular saws and band saws are both employed.

Flame cutting is, unfortunately, no good for aluminium because of the ragged edge produced. For profiled cuts, it is necessary to use band sawing or plasma cutting.

3.1.3 Holing

Drilling of aluminium is faster than with steel, and is the normal method of making holes. When punching is used, it is desirable to punch to 75% of the final hole size and drill out. This applies especially when the diameter is less than the thickness, leading to enlargement of the hole on the tear-out side. The practice of predrilling holes for fasteners to a smaller diameter, and reaming after assembly, is commoner in aluminium than in steel.

Hole spacing and edge distances must not be made too small, so as to avoid premature tearing in joints loaded in shear. British Standard

BS.8118 imposes the following limits, measured to hole centres: hole spacing $\geq 2.5d_0$; edge distance $\geq 1.5d_0$, where d_0 is the nominal hole diameter. This minimum value for edge distance is safe if applied to an extruded edge, but with a cut edge (sawn or sheared) it may be prudent to increase it to allow for inaccurate cutting.

3.1.4 Forming

Procedures for bending and forming aluminium are broadly similar to those for structural steel, and the same equipment is generally suitable. However, some aluminium materials are less readily formed than steel, especially heat-treated alloys in the full-strength T6 condition. The formability of non-heat-treatable material depends on the temper in which it is supplied; it is readily bent in the softer tempers. Data on the formability of different alloys appears in Chapter 4 (Section 4.3.4). Because of aluminium's lower elastic modulus (E), springback is greater than with steel.

Heat-treated material should only be manipulated when cold, and bending it in the full-strength T6 condition is difficult. The 6xxx-series alloys, and even worse the 2xxx alloys, will accept very limited deformation in this condition, and such practice is forbidden for 7xxx-series material, because of the risk of stress-corrosion cracking. Ideally, heat-treatable material should be bent in the T4 condition soon after solution treatment, before it has had time to age-harden. The maximum time after quenching is about two hours, although this can be extended to four or five days by cold storage (−6 to −10°C). A degree of deformation is still possible if carried out later, i.e. after the metal has undergone natural ageing and reached its T4 properties. In either case, the material can be brought up to full T6 properties by subsequent precipitation treatment (artificial ageing).

Such procedures are possible only if suitable heat-treatment facilities exist. Solution treatment involves holding the part at a temperature slightly above 500°C for 20 or 30 minutes and then quenching it in water, which causes distortion. The temperature needed for precipitation treatment is lower (under 180°C) and less critical, the required holding time being several hours.

The forming of work-hardened (non-heat-treatable) material can be facilitated by applying local heat at the bend. A problem in so doing is the difficulty in controlling the degree of heat, because unlike steel aluminium does not change colour when the right temperature is reached. This can be overcome in several ways. One way is to apply a film of soap and see when this turns black. Another method is to go on heating until a pine stick scraped along the surface leaves a char mark. A more accurate technique is to employ temperature-indicating crayons. Minimum annealing temperatures are: 1xxx-series, 360°C; 3xxx-series, 400°C; 5xxx-series, 350°C.

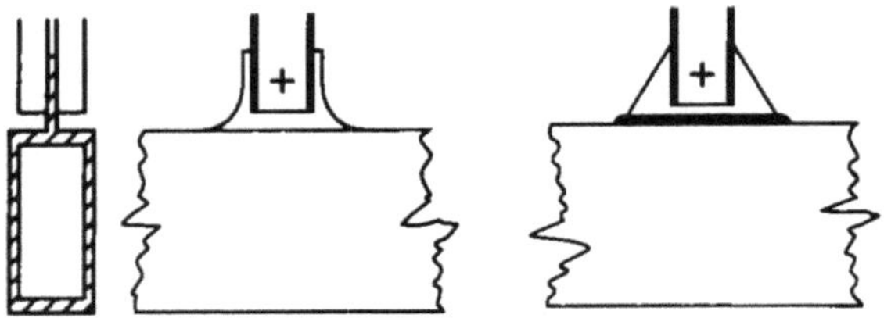

Figure 3.1 Alternative designs for a gusset attachment: (a) machined; (b) welded.

3.1.5 Machining

Because of the high metal removal rate that is possible with fully-heat-treated material in the strong alloys, a valid technique for the manufacture of large structural components is to machine them. For the aircraft of World War II, this was the standard way of producing spar-booms (extending the length of a wing), these being machined out of thick extrusions. Since then the method has been extended to the milling of wide stiffened panels out of thick plate. Such panels, forming large parts of a wing or fuselage, typically comprise a skin of varying thickness with integral stiffeners, and are machined in the flat before curving. Large computer-controlled milling machines have been developed for such manufacture, in which 80% or more of the original plate may be removed.

Two operations must be carried out on such material before machining begins. One is to stress-relieve it and so prevent distortion as metal is removed. This is done by stretching, the required force in the case of thick plates being several thousand tonnes. The other essential operation is ultrasonic inspection, to check for the possible presence of small inclusions in the aluminium.

It is not only in the aero-industry that machining is a valid form of fabrication for structural components. A simple example would be the chord section shown in Figure 3.1 to which a cross-member has to be bolted. In design (a) an extrusion is used, with an integral flange that is machined away over most of its length. The alternative method of welding on a local gusset (b) would reduce the strength of the member, especially in fatigue.

3.2 MECHANICAL JOINTS

3.2.1 Bolting and screwing

Bolts in aluminium structures can be *close-fitting*, in reamed holes, or *clearance*. For the latter, BS.8118 originally limited the clearance to 0.4 mm or 0.8 mm respectively, for diameters below and above 13 mm. This has now been amended (1997) to 1.6 mm for all sizes, which is more in line with other codes.

The main decision in the design of a bolted joint is the choice of bolt material, for which there are three basic possibilities:

- aluminium bolts;
- austenitic stainless steel bolts;
- steel bolts (suitably protected).

It is penny-pinching to select a bolt material that is less corrosion-resistant than the aluminium parts being connected. Aluminium and stainless steel are the preferred materials in this respect.

Aluminium bolts for non-aeronautical use are usually in the stronger kind of 6xxx-series alloy. They can be either machined from T6-condition bar-stock, or forged and then artificially aged (T8-condition). Design data for such bolts is included in Chapter 11 (Table 11.1). Also available are bolts in 2xxx-series material, which are stronger, though these may pose corrosion problems. It is recommended that the threads of aluminium bolts be lubricated, especially if the joint is later going to be dismantled.

The best bolts for aluminium structures are in austenitic stainless steel (300-series), as these are much stronger than aluminium ones. The extra cost is likely to be small relative to that of the final structure. British Standard BS.8118 permits the use of A4 or A2 stainless bolts (see BS.6105), each of which comes in three possible grades with tensile strengths of 500, 700 and 800 N/mm^2 respectively.

Steel bolts (non-stainless), which must have a protective coating, can only be considered for indoor use. When exposed to the elements it is simply a matter of time before the coating disappears and they begin to rust. The bolts can be cadmium-plated (or equivalent), or else have a more durable coating of zinc, as obtained by galvanizing or sherardizing. When there is a risk of plated steel bolts getting muddled up with a batch of stainless steel ones, they can be separated magnetically, the stainless being non-magnetic (if of the austenitic type).

British Standard BS.8118 calls for washers to be provided under both bolt-head and nut. With aluminium and stainless steel bolts, these would normally be of aluminium, in a comparable alloy to the parts being joined. However, the British Standard also allows washers of pure aluminium.

In specifying the length of thread and the thickness of washers, it is important to consider whether or not threads are allowed to come within the thickness of a part, since this affects the resistance of the joint to failure in bearing (Section 11.1.4).

The use of set-screws with aluminium components is not generally favoured, because of the unreliability of threads in aluminium. When such a detail cannot be avoided, best practice is tap out the hole to a larger size, and introduce a *thread insert* (Figure 3.2). This is a device resembling

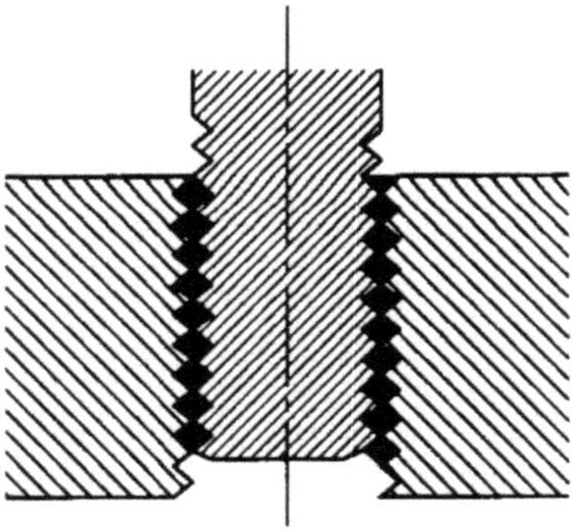

Figure 3.2 Thread insert.

a coil-spring, made of rhomboid section stainless steel wire, which can be screwed into the hole. It greatly increases the life of the thread if the joint has to be unscrewed from time to time.

3.2.2 Friction-grip bolting

This type of connection is employed when maximum rigidity is needed under shear loading. It calls for the use of special high tensile (HSFG) steel bolts in clearance holes, these being torqued up to a high tension so that the service loading is carried entirely by friction. The technique is less advantageous than in steel (Section 11.2), but it still has some application. Practice broadly follows that for steel structures.

The bolts should be of *general grade* (BS.4395: Part 1). HSFG bolts of higher grade are held to be unsuitable for aluminium, and BS.8118 restricts even the general grade to use with aluminium having a proof stress over 230 N/mm^2.

It is important that the mating surfaces should be flat, clean, free of defects, and also free from any substance that would reduce the development of friction, such as paint. Preferably, the surfaces should be grit-blasted. British Standard BS.8118 recommends a standard treatment, and when this is used a stated design value may be taken for the slip-factor (Section 11.2.6).

Tightening procedures are as for joints in steel, either by controlled torque using a calibrated wrench, or else using the part-turn method. Refer to BS.4604.

3.2.3 Riveting

In light-gauge construction, the decision whether to rivet or weld is slanted less towards welding than it is in steel. Small aluminium rivets are widely used in some industries, either of conventional solid form or else of ingenious 'proprietary' design. In the late 1940s, large aluminium rivets came into use (up to 25 mm diameter), and some notable structures were built with them. High forces were needed to close these rivets and special squeeze-riveters were developed for the purpose. All this disappeared with the arrival of gas-shielded arc welding in the 1950s, combined with the fact that riveting had become a lost art generally in the fabricating industry. But one wonders whether the rush to weld aluminium has not been somewhat Gadarene.

Non-heat-treatable solid aluminium rivets come in 5xxx-series alloy. For the best shear strength, they should be supplied in a work-hardened condition (typically quarter-hard) and driven cold. To facilitate closing (i.e. for larger rivets), they can instead be used in the annealed O-condition, with a slight strength penalty. In this case, they can if necessary be driven hot, the required temperature being about 350°C. Table 11.1, based on BS.8118, covers two typical rivet materials of this kind, 5154A and 5056A. The latter, which is the stronger, is not recommended for tropical environments because of possible corrosion.

The alternative is to employ rivets in heat-treatable material. These are driven cold, normally in the solution-treated T4 condition. For the greatest ease of closing, they should be used within two hours of quenching, or up to four or five days if held in a refrigerator (at –6 to –10°C). Rivets of this kind would normally be in the stronger kind of 6xxx-series alloy, such as 6082, for which strength data are included in Table 11.1. The British Standard also includes data for rivets in the T6 condition.

Aluminium rivets should always be used in reamed holes, as they will not tolerate a bad hole in the way that hot-driven steel rivets used to. Because they are driven cold, or at a fairly modest temperature, they exert negligible clamping action between the plates. On the other hand, after driving, they fill the hole well. Aluminium rivets should not be used in situations where they have to carry tensile loading.

The above information refers to conventional solid rivets. In the attachment of sheet metal panels much use is made of proprietary fasteners, which are easy to use and which are suitable for blind riveting (access to only side of the joint). Examples of these are the well-known 'pop' and 'chobert' rivets, both of tubular form. Pop-rivets come in diameters up to about 5 mm. They can exert a limited clamping action, provided care is taken with plate fit-up during closing, but they do not fill the hole so well as a cold-driven solid rivet. In contrast, the chobert rivet, which is available in larger sizes, fills the hole well but has negligible clamping action. A wide range of other proprietary fasteners is available, generally more expensive, mostly for use in the aero-industry. In the USA such fasteners, which are a cross between a bolt and a rivet, are also popular in non-aeronautical construction.

3.3 ARC WELDING

3.3.1 Use of arc welding

Over the last 25 years, arc welding has achieved complete acceptance as a method for joining aluminium, following the American development of *gas-shielded* welding in the 1950s. These replaced ordinary stick welding, which had proved useless for aluminium. Two gas-shielded processes are available: MIG (Metal Inert Gas) and TIG (Tungsten Inert Gas). MIG is the more widely employed, especially for heavier construction, using weld geometries and preparations similar to those in structural steel. TIG is employed for small welds in light gauge material, and also for making repairs to MIG welds.

The essential feature in both processes is a flow of inert gas from the welding torch, which shields the arc and weld-pool from the air and so prevents oxidation. The standard gas for aluminium is argon. An alternative possibility is helium, although this has limited application today. For outdoor welding, it is necessary to provide a booth, to stop the gas blowing away in the wind.

Gas-shielded arc welding can be used with most kinds of aluminium, including 1xxx, 3xxx, 5xxx and 6xxx-series materials. It is no good for the 2xxx alloys because of cracking, nor for the stronger kind of 7xxx material. The weaker 7xxx-series alloys are weldable, but with satisfactory results depending critically on good technique.

Welded joints in aluminium, unlike structural steel, tend to suffer serious weakening in the heat affected zone (HAZ) surrounding the weld, known as *HAZ softening* (see Chapter 6). The strength in this zone can be nearly halved, as for example with fully-heat-treated 6xxx material (T6 condition), and the weakening may easily extend 25 mm or more out from the weld. With the 7xxx-series alloys, the weakening is less severe, but extends further. When tested to destruction, aluminium joints sometimes fail in the weld metal and sometimes in the HAZ, depending on the combination of parent and filler alloys being used.

In making large multi-pass welds, the fabricator should exercize thermal control, to prevent excessive build-up of heat and consequent enlargement of the softened zone. This is achieved by monitoring the interpass temperature (Section 6.2). Despite the need to limit the extent of the HAZ, it is sometimes desirable to apply preheat. Typically this would be used when welding thin to thick, in order to achieve proper side-wall fusion.

3.3.2 MIG welding

MIG is a direct current (DC) process with electrode positive. It is similar to CO_2 welding of steel. The electrode is in the form of wire, supplied in a coil and of typical diameter about 1.5 mm. It is fed automatically through the torch and into the arc while the weld is being made. The torch, which is water-cooled, is much more complicated than that used for ordinary stick welding, since the supply tube has to convey four things simultaneously: current, wire, argon and water. The MIG process is suitable for welds in material down to a minimum thickness of 4 or 5 mm.

Normally the electrode wire is pushed down the tube leading to the torch. This can occasionally cause problems with the wire kinking and jamming in the tube. The equipment for making the smallest MIG welds, using 0.8 mm wire, avoids this problem by pulling the wire through the tube. This adds further to the complexity of the torch.

The MIG process has two special attributes. First, it is suitable for positional welding, including overhead. Second, the arc-length is self-adjusting. In other words, when the welder accidentally moves the torch nearer or further away from the job, the burn-off rate momentarily changes until the right amount of wire is sticking out again, thus maintaining the correct arc-length automatically. This 'semi-automatic' feature makes it a relatively easy process to use.

When welding in the downhand position, the section area of the maximum possible size of MIG deposit is some 40 mm^2 per pass. Much higher welding speeds are possible than with stick welding of steel. For a weld on 6 mm plate, a speed of 1.5 m/s would be reasonable, compared with, say, 0.5 m/s on steel.

3.3.3 TIG welding

TIG is an alternating current (AC) process. The water-cooled torch has a non-consumable tungsten electrode, and as with MIG it delivers a flow of argon. The filler wire is held in the other hand and fed in separately. TIG requires more skill than MIG, both to control the arc length and to feed in the filler. One disadvantage is that the operator may inadvertently let the torch dwell in one position, causing a local build-up of heat and hence an enlarged HAZ.

The TIG process is suitable for the welding of sheet thicknesses, rather than plate. At the top end of its thickness range (say 6 mm), the penetration can be improved by using a helium shield instead of argon.

It is possible to adapt the TIG process to *autogenous* welding, i.e. dispense with the use of filler wire. The necessary filler metal is provided by ribs on the parts being joined, which melt into the pool. Such a technique enables TIG to be set up for automatic welding, with machine-controlled traversing of the torch.

A useful variant of TIG is the *pulsed-arc* process. This employs a DC supply with current modulation, causing the weld to be produced as a series of nuggets, which fuse together if the parameters are suitably adjusted. The technique is claimed to be superior to ordinary TIG when welding thin sheet, where it provides easier control of penetration. Also there is reduced build-up of heat.

3.3.4 Filler metal

Specification of the electrode/filler wire for MIG or TIG is a design decision and should not be left to the fabricator. It is mainly a function of the alloy of the parts being joined, known as the 'parent metal'. Sometimes there is a choice between more than one possible filler alloy, and the selection then depends on which of the following factors is the most important: weld metal strength; corrosion resistance; or crack prevention.

We divide possible filler alloys into four types, numbered to correspond with the alloy series to which they belong:

Type 1	pure aluminium
Type 3	Al–Mn
Type 4	Al–Si
Type 5	Al–Mg

Table 3.1 (based on BS.8118) indicates the appropriate filler type to select, depending on the parent alloy. Compositions of actual filler alloys within each type are given in Chapter 4 (Table 4.8). When the parts to be connected are of differing alloy type, the choice of filler may be arrived at with the aid of Table 3.2, which is again based on the British Standard.

Type 1 fillers

The purity of the filler wire should match that of the parent metal.

Table 3.1 Selection of filler wire type for MIG and TIG welding

Parent alloy series	Filler alloy	Comments
1xxx	Type 1 or 4	Type 1 for corrosion resistance. Type 4 for strength and crack prevention.
3xxx	Type 3 or 4	Type 3 for corrosion resistance. Type 4 for strength and crack prevention.
5xxx	Type 5	Important to select *appropriate* Type 5 composition, depending on actual parent composition.
6xxx	Type 4 or 5	Type 4 for crack prevention. Type 5 may give better weld strength.
7xxx	Type 5	When strength is paramount, it is important to select *appropriate* Type 5 filler composition.

Type 4 fillers

The advantage of these, when used for welding 1xxx, 3xxx or 6xxx-series material, is their ability to prevent cracking. The normally preferred version is 4043A (≈5% Si). When crack control is paramount, as in a highly restrained joint, the 4047A (≈12% Si) filler may be specified instead, but with some loss in corrosion resistance. Type 4 fillers provide no protection against cracking when employed for welding parent metal containing over 2% Mg, and are therefore unsuitable for use with most 5xxx and 7xxx materials.

Type 5 fillers

When a joint is to be made in 7xxx-series alloy or the strongest form of 5xxx (such as 5083), and the prime requirement is weld-metal strength, the best filler alloy is 5556A (or equivalent). For other 5xxx-series parent alloys, the precise choice of filler tends to be less critical. As a general rule, the filler composition should broadly match that of the parent metal. A lower amount of magnesium will produce a weaker weld, whereas a more highly alloyed filler may lead to trouble in potentially corrosive environments.

3.3.5 Weld inspection

The required amount of inspection at a welded joint depends on the level of weld quality that the designer wants to achieve, and BS.8118 recognizes three such levels [9].

1. *Minimum quality*. This is acceptable when the force transmitted by the joint under factored loading is not more than one-third of its factored static resistance, and fatigue is not a factor.
2. *Normal quality* is called for in non-fatigue situations when the transmitted force is too high for minimum quality to be allowed, or under fatigue conditions when the required fatigue class is 20 or below.
3. *Fatigue quality* applies when fatigue is a factor in the design of the joint, and the required fatigue class is 24 or over.

Table 3.2 Filler wire types for welding dissimilar parent alloys

	Alloy-series of second parent metal				
Alloy-series of first parent metal	1xxx	3xxx	5xxx	6xxx	7xxx
1xxx	1 or 4	3 or 4	5	4	5
3xxx	3 or 4	3 or 4	5	4	5
5xxx	5	5	5	5	5
6xxx	4	4	5	4 or 5	5
7xxx	5	5	5	5	5

For welds of minimum or fatigue quality, the designer must indicate the quality level on the drawing. Where no such information is given, the fabricator assumes that normal quality applies. For fatigue quality joints the drawing should include the *Fat-number* and also an arrow showing the direction of the fluctuating stress. As explained in Section 12.7, the Fat-number defines the required fatigue class (24, 29, 35, 42 or 50), which may be less than the highest class possible for the detail concerned.

The following is a summary of the main types of inspection method that are available for aluminium welds:

1. Visual inspection before welding (surface condition, preparation, alignment, fit-up, etc);
2. Visual inspection after welding (weld geometry, profile discontinuities, alignment, etc);
3. Non-destructive testing (NDT), which aims at checking all surface and sub-surface defects (cracks, porosity, inclusions etc):
   - dye penetrant testing;
   - ultrasonic testing;
   - radiography.
4. Taking macro-sections from separate test-pieces, which are welded at the same time and properly represent the technique and conditions in the production joint.

British Standard BS.8118 provides comprehensive tables stating which inspection methods are needed when, and also the recommended acceptance levels for different kinds of defect. The amount of inspection called for depends on the required quality level, the type and orientation of the weld, and the Fat-number (when relevant). Minimum quality welds only have to be visually inspected. Dye penetrant tests are generally required for normal or fatigue quality, with ultrasonics or radiography being necessary as well at transversely loaded butts. Macros are required only for fatigue-quality welds of higher Fat-number.

3.4 FRICTION-STIR WELDING

3.4.1 The process

The friction-stir (FS) process is a solid-phase welding process, invented in 1991 at the TWI (formerly The Welding Institute), near Cambridge, UK. [10]. It has been developed in the first place for welding aluminium, where it is set to become an accepted technique along with MIG and TIG, and at the time of writing it is being exploited commercially in several countries [11]. Possible extension of the FS process to steel poses a

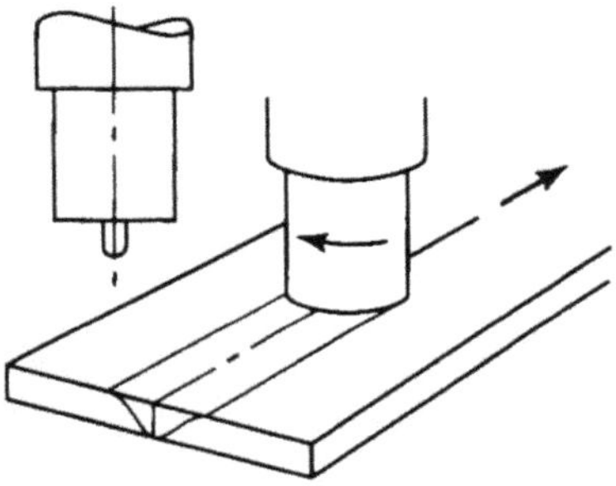

Figure 3.3 Friction-stir welding: the process.

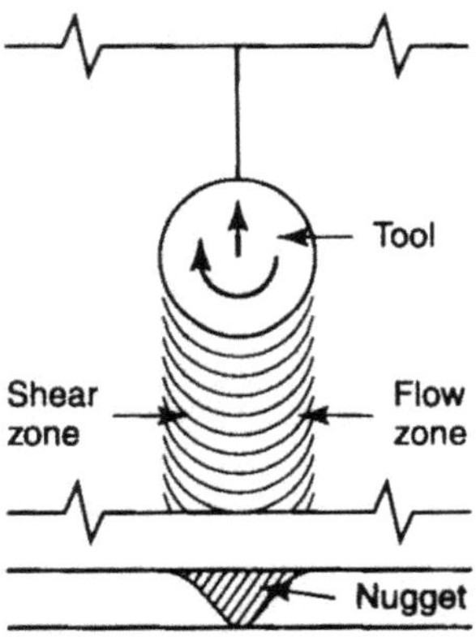

Figure 3.4 Friction-stir welding: typical scratch marks on top surface and shape of 'nugget'.

problem, because of the need for a tool material able to withstand the much higher welding temperature.

Figure 3.3 shows how FS works. The plates to be joined are butted together on top of a firm support bar, and are clamped so as not to move apart during welding. The process employs a steel welding tool comprising a pin and a shoulder, in which the height of the pin is made slightly less than the plate thickness. The rotating tool is plunged into the aluminium plates until the shoulder meets the top surface, and is then traversed along the line of the weld. Frictional heating under the shoulder causes a zone of aluminium to become plastic. Violent 'mashing' and 'stirring' occurs in this zone and the plates weld together, by a mechanism not unlike the formation of seam-welds in hollow extrusions.

A joint made by FS welding is flat on both sides, with a characteristic scratch pattern on the top surface as depicted in Figure 3.4 ('onion-ring' structure). The figure also shows the typical cross-section of the 'nugget' of heavily worked material, comprising a 'shear zone' and a 'flow zone'.

The process has been successfully used for welding plates from 3 to 12 mm thick. The welding tool is in hard heat-resisting steel, with a typical

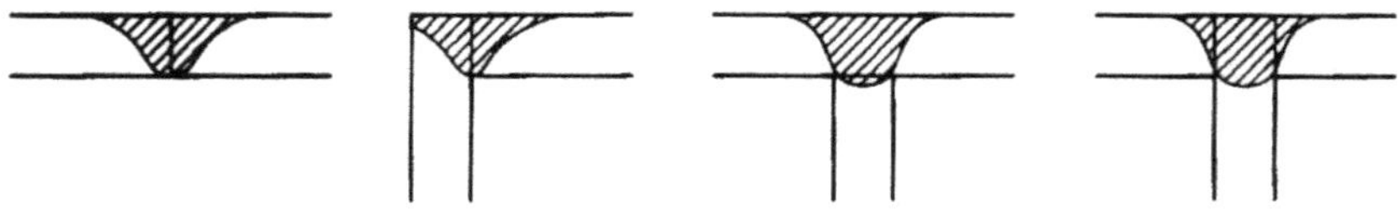

Figure 3.5 Friction-stir welding: some joint geometries.

shoulder diameter in the range 10–20 mm. The Norwegian firm Hydro Aluminium a.s. has patented a modified form of tool in which the pin has two circumferential ribs and the shoulder is slightly dished.

The force exerted on the plates by the shoulder is an important parameter which must be monitored and controlled during welding. The other welding parameters are shoulder diameter, tool rpm and tool traverse speed. Figure 3.5 indicates some of the joint geometries that can be accommodated.

3.4.2 Features of FS welding

FS welding provides a simple means of making butt-joints, for which it has the following advantages over MIG and TIG.

- *Profile*. An FS joint, with its flat surfaces, has a tidier geometry than an arc-welded joint.
- *Strength*. From data so far available, it appears that FS joints are stronger than arc-welded ones. The mechanical properties in the nugget have been found to be superior to those in the HAZ at an arc weld; also the question of weld-metal strength (i.e. the deposit) does not arise, because there is none.
- *Fatigue*.Clearly FS joints are superior in fatigue, because of their good profile.
- *Grain size*. The nugget in FS welds has a finer grain size than that of the parent metal, causing them to perform well in transverse bend tests.
- *Quality control*. Once the welding parameters have been established, an FS weld is less operator dependent.
- *Weldable alloys*. The FS process is suitable for welding all the main alloy types, including the 2xxx-series, and the strong kind of 7xxx (such as 7075).
- *Equipment* for the FS process is relatively unsophisticated.
- *Power consumption* is less.
- *Weld-preparation*. The plates are simply butted, without any preparation.

3.4.3 Limitations

The chief disadvantage of the FS process is the need to provide a rigid support system, to react against the considerable downward force exerted by the welding tool. This calls for the design of special FS welding machines, either with the job moving past a fixed welding head, or else with the job stationary and the welding head moving, the latter arrangement being necessary for making very long welds. (Early experiments made use of a suitably modified milling machine, the travel on which limited the possible length of weld.)

Another snag is that a hole remains at the end of each weld run, formed when the tool is withdrawn. This has to be filled in somehow, an available method being *friction taper plug* welding. As with arc welds, run-on and run-off plates are generally necessary, unless it is possible to cut off the end material.

3.4.4 Applications

A major role for the FS process is likely to be in the manufacture of large areas of stiffened plating, by welding together long lengths of extruded planking. Such a product is attractive in truck bodies, ships, offshore modules and bridges. The equipment can also be installed in extrusion plants, where it provides an economic means of supplying wide sections by welding together narrower ones, thus avoiding the high capital cost of a very large press. Such a technique can be employed for making *double-skin* stiffened plating, by welding together plank sections of rectangular hollow profile, using an FS weld on either surface.

3.5 OTHER WELDING PROCESSES

Resistance welds are readily made in aluminium, generally with a higher energy input than with steel. Spot welding of car bodies is standard, latterly in conjunction with adhesive bonding. But flash-butt welding of window frames has entirely given way to mechanical joints. Other forms of friction welding (beside friction-stir) are also available.

3.6 ADHESIVE BONDING

3.6.1 Use of bonding

As a connection technique for aluminium, adhesive bonding has features that often make it a valid alternative to bolting, riveting or welding (Section 11.4.1). But its successful use is critically dependent on good practice by the fabricator, especially as regards preparation of the surfaces.

Adhesives used with aluminium are usually epoxy materials, two main types being available: the two-component and the one-component. None are covered by official standards in the way that metal products are, and the designer/fabricator must rely closely on information provided by the adhesive maker. Tables 11.7 and 11.8 in Chapter 11 provide data on selected adhesives supplied by one particular manufacturer (Ciba).

3.6.2 Surface preparation

The success of a bonded joint stands or falls on the care with which the mating surfaces are prepared. Possible levels of such preparation in increasing order of efficacy (and cost) are:

1. degrease only;
2. degrease, abrade and remove loose particles;
3. degrease and pretreat.

The appropriate level must be specified by the designer. The object of using level (2), or best of all (3), instead of (1), is to make the joint more reliable and durable in humid environments.

(a) Degreasing

The prime requirement in preparing surfaces for bonding is to remove all traces of oil and grease, even though the surfaces may appear clean. The following are acceptable degreasing methods.

- *Vapour degreasing*. The parts are suspended in the vapour of a halocarbon solvent, such as trichloroethylene ('trike'). This calls for a special vapour degreasing unit.
- *Immersion degreasing*. The parts are successively immersed in two tanks of the same solvent in liquid form, the first tank acting as a wash and the second as a rinse.
- *Wiping*. The surfaces are degreased using a clean brush or cloth soaked in halocarbon or other degreasing solvent.
- *Detergent degreasing*. The surfaces are scrubbed in a solution of liquid detergent, and then washed with clean hot water, followed by drying in air.

A simple test to find whether a part has been properly degreased is to see if distilled water wets the surface and spreads, rather than forms droplets.

(b) Abrading

Light abrasion of the surfaces after degreasing gives a better key to the adhesive. The preferred method is to grit blast, after which all loose particles must be removed, using a clean soft brush or compressed air. Another way to remove loose bits is to repeat the degrease operation.

If grit-blast equipment is unavailable, or the part is too thin, abrasion can be carried out using a wire brush, abrasive cloth or a water-proof

abrasive paper. With these, the subsequent removal of contaminants and dust is facilitated by the use of water, followed by drying (in air).

(c) Pretreatment

Following the degrease a recommended procedure, better than abrading, is to etch the surfaces in a solution of sulphuric acid and sodium dichromate. This must be followed by washing in water and drying in air. Another possible procedure is anodizing (Section 3.7.3), after which no further treatment is needed. Note, however, that a *hard-anodized* surface is not recommended as a pretreatment for bonding.

3.6.3 Two-component adhesives

These are supplied with the resin and hardener in separate containers. Curing begins as soon as they have been mixed, i.e. just before making the joint. There are then two possibilities. Either the joint can be held at room temperature, in which case it will take some hours or possibly a day to cure ('go off') and develop its strength. Alternatively it can be cured under heat, enabling full strength to be attained much more quickly.

The two-component adhesives are used when the assembled parts can be safely left for some time before handling, while curing takes place, or when the parts cannot be heated for some reason.

3.6.4 One-component adhesives

With the one-component adhesives, the resin and hardener come ready mixed in a single container, chemical reaction between the two being inhibited as long as they stay at room temperature. After application of the adhesive, the joint is cured by heating the assembled component in an oven. The heating cycle can typically range from one hour at 120°C to 10 minutes at 180°C.

The one-component adhesives tend to be stronger than the two-component. They are selected when the bonding/curing cycle has to be reasonably quick, to fit into the overall production schedule.

3.6.5 Applying the adhesive

After the surfaces have been prepared, it is essential to prevent any subsequent contamination, even touching with a finger. The adhesive should be applied as soon as is reasonably possible, but not before the surfaces have had time to dry. It need only be applied to one surface.

The adhesives come in a range of different pack types and sizes. A convenient method is to use special cartridges that can be loaded into a dispensing gun. The gun for two-component adhesives accepts a double cartridge, and mixing takes place automatically in the correct ratio as the gun is operated, by means of a special nozzle. On some jobs, it may be more convenient to employ a spatula, for both mixing and application. After use, equipment should be cleaned with hot water and soap before residues have had time to cure and harden.

3.6.6 Clamping

It is obviously important to prevent any relative movement of the parts until an adequate degree of curing has taken place, the time involved being a function of the adhesive used (Section 11.4.10). The necessary clamping can often be simplified by suitable design, as for example by using a tongue-and-groove detail between mating extrusions. The clamping technique may be important in controlling the glue-line thickness.

3.6.7 Curing

Two-component adhesives can be satisfactorily cured at room temperature, although this tends to be slow. For a faster cure, the assembly may be warmed, typically in the range 40–80°C, and this has the added advantage of increasing the strength. A radiant heater, electric blanket or hot-air blower may be employed for the warming, which can be controlled with the use of temperature-sensitive paper. Refer to Figure 11.17 for the variation of cure time with temperature for six selected adhesives.

One-component adhesives are cured by heating the assembled job in an oven, the necessary time being a function of the temperature, which typically lies in the range 120 to 180°C.

3.7 PROTECTION AND FINISHING

3.7.1 General description

The reason for specifying a surface finish on aluminium is either protection against corrosion, or else appearance. Possible finishes are anodizing (Section 3.7.3) and painting (Section 3.7.4). There are two essential ways in which aluminium can corrode, both of which are discussed in Chapter 4 (Section 4.7):

1. *General corrosion of an exposed surface*. In the great majority of installations, an aluminium surface can be left unpainted, in its as-supplied *mill-finish* state. Such corrosion as may occur will not threaten the structure. Situations that demand a protective finish are defined in Section 4.7.2.
2. *Corrosion due to contact with certain other materials*. This may lead to serious attack on the aluminium and early failure. Suitable precautions are essential.

3.7.2 Pretreatment

Surfaces that are going to be anodized or painted should receive a suitable pretreatment, which normally comprises three successive dips: degrease, rinse and etch. Many extruders offer such a pretreatment service. Various formulations are employed for the etching agent, and these have an effect on the final appearance when a component is later anodized.

3.7.3 Anodizing

Anodizing is a process in which the thin oxide layer, always present on an aluminium surface, is artificially increased. The effect is to improve the corrosion resistance and also the appearance, as compared with mill finish. The anodized surface can be natural colour ('satin' finish) or coloured.

Anodizing is a low-voltage high-current electrolytic process whereby the component is immersed in an acid electrolyte (usually sulphuric), through which direct current is passed with the aluminium as anode. The relatively thick oxide film thus produced contains capillary pores. The final operation is to seal these pores by placing the component in boiling water, which produces a hard weather-resisting surface. Colour can be introduced by immersion in a dye tank, before sealing.

The appearance of an anodized surface is affected by the alloy-group of the aluminium, and also by the details of the pretreatment. In welded items, the appearance of the welds will be clearly different from that of the parent metal, if the filler wire is of different composition as it usually is.

The thickness of the oxide film on anodized components should be specified by the designer. For indoor applications, where appearance is all that matters, the thickness would normally lie in the range from 0.010 to 0.015 mm (some 2 to 3 times the natural value). For outdoor use, where protection and durability of the surface are needed, the recommended minimum is 0.025 mm, known as 'architectural' grade anodizing.

It is possible to use chromic acid for the electolyte, instead of the usual sulphuric. The resultant film is less hard and its maximum possible thickness is only 0.01 mm. But, for a given film thickness, the chromic process provides better corrosion resistance, making it a valid alternative for aggressive industrial environments. It is recommended for components containing crevices that might retain electrolyte, even after rinsing, since aluminium is not attacked by chromic acid.

Another version of the process is *hard anodizing*, which enables the oxide film thickness to be increased to 0.125 mm. Such a film is dense and hard, and relatively rough. It is normally left unsealed, but can be waxed or treated with mineral oil. Hard anodizing is mainly specified for its high resistance to abrasion, which compares favourably with that of chromium plating on steel. Corrosion resistance is excellent. The film also acts as an electric insulator.

3.7.4 Painting

Alloys having durability rating A or B do not normally need to be painted, except for the sake of appearance (Section 4.7.2). A paint finish is more likely to be called for with C and D-rated alloys, to provide protection. A brief outline of possible paint systems is given below. Whatever system is used, a suitable pretreatment is essential.

Mass-produced components can be painted in the factory using *powder coatings* which have largely replaced solvent-based paints for such application. Before powder coating, the pretreated items must first receive a *chemical conversion* coating. They then pass through an electrostatic field, where they are subjected to a spray of dry powdered paint material, after which they proceed through a tunnel oven for stoving. The stoving cycle is typically of the order of 200°C for 10 minutes. A feature of the process is its lack of waste, since any excess sprayed powder falls to the floor where it can be collected and re-used.

For non-factory application, aluminium can be coated in the normal way using solvent-based paints. It is important that none of these, especially the priming coat, should contain copper, mercury, tin, lead or carbon-based compounds. A preferred primer is one based on zinc chromate.

3.7.5 Contact with other materials

At joints between aluminium and most other common metals, where moisture may be present, it is essential to take suitable precautions to prevent bimetallic corrosion (Section 4.7.3). British Standard BS.8118 gives detailed advice on the treatment of such joints. The main essential is to insulate the two surfaces from each other using, in order of efficacy, a coat of priming paint, an impervious tape, or a non-conducting gasket. The other requirement is to apply comparable treatment to bolt holes and under bolt heads, and to protect the heads of steel bolts with aluminium paint, or by metal spraying with aluminium.

There are occasions when aluminium needs to be protected from non-metallic materials with which it is in contact, in the presence of moisture. Such materials are concrete (while still 'green'), soil, some woods (e.g. oak) and some insulation boards. Treatment with bituminous paint is usually efficacious in such situations.

CHAPTER 4

Aluminium alloys and their properties

4.1 NUMBERING SYSTEM FOR WROUGHT ALLOYS

4.1.1 Basic system

The American system for designating wrought aluminium materials, administered by the Aluminum Association (AA) in Washington, is by now virtually standard worldwide [12]. For any given material, it employs a two-part reference number, e.g. 3103-H14, 6082-T6 or 5083-0. The four-figure number before the hyphen defines the alloy, and the symbols after it the condition (or temper) in which that alloy is supplied.

Aluminium alloys can be either non-heat-treatable or heat-treatable. The former are available in a range of work-hardened conditions, defined by an H-number (as in the first example above), while the latter are supplied in various conditions of heat treatment, specified in terms of a T-number (second example).

4.1.2 Standardization of alloys

The four-figure number before the hyphen relates to a chemical composition, i.e. an alloy. The AA supply a document (*Registration Record of International Alloy Designations*) which lists the reference numbers of all the alloys registered with them, together with their compositions. The list runs to several hundred and is regularly updated. The new European (EN) standards have adopted the same alloy numbers, but prefixed by the letters AW (Aluminium Wrought). Thus, in the European system, 6082 alloy is officially referred to as AW-6082, although in common speech it is normal to omit the AW.

The following illustrates the way in which a composition is specified, using the alloy 6082 as an example:

Si	Mn	Mg	Fe	Cu	Cr	Zn	
0.7–1.3	0.4–1.0	0.6–1.2	0.5	0.1	0.25	0.2	%

Where a range is given, as for the first three elements, this denotes the minimum and maximum permitted content of an *intended* ingredient. Where a single figure is given (as for the last four) this refers to an *impurity*, which must not exceed the value quoted. This is the official way of specifying such an alloy. In everyday parlance, it is more convenient to talk in terms of the *nominal* composition, quoting the mean percentages of the intended elements only. The above 6082 alloy would be described as containing a nominal:

Si 1.0 Mn 0.7 Mg 0.9%

Aluminium alloys are grouped into seven *alloy series*, depending on their main alloying ingredient(s) (Table 4.1). The first digit of an alloy number indicates the series to which that alloy belongs, while the other three usually have no particular significance. An alloy series may be referred to collectively by putting xxx after the first digit, i.e. the '5xxx series'.

Sometimes a slightly modified version of an existing alloy is registered, in which case an 'A' may be added to distinguish it from the original (e.g. 5154 and 5154A).

The AA also carry an 8xxx series, used for compositions that do not easily fit into any of the first seven. One such type of alloy, containing a substantial amount of lithium, has raised interest in the aircraft industry because of its significantly lower density compared to other aluminium alloys.

4.1.3 Work hardening

Non-heat-treatable alloys are strengthened by means of cold work applied during manufacture, as in the cold reduction of sheet or drawn tube. The reduction per pass, and any heating, are controlled to produce the required combination of strength and ductility. For a given degree of cold work, the resulting increase in the proof stress (yield) is relatively much greater than the increase in tensile strength (UTS), as compared with the annealed condition.

The condition (or temper) of work-hardened sheet and drawn tube can be crudely expressed in terms of its hardness, the standard tempers being

Table 4.1 The seven alloy series

Alloy series	Main alloying ingredient(s)	Type
1xxx	(Pure aluminium)	NHT
2xxx	Copper, Cu	HT
3xxx	Manganese, Mn	NHT
4xxx	Silicon, Si	NHT
5xxx	Magnesium, Mg	NHT
6xxx	Magnesium and silicon, MgSi	HT
7xxx	Zinc and magnesium, ZnMg	HT

Note. NHT = non-heat-treatable, HT = heat-treatable

Table 4.2 Temper designations for non-heat-treatable material

Condition	Temper designation		
	Temper-rolled	Temper-annealed	Stabilized
'Fully hard'	H18	H28	H38
'3/4 hard'	H16	H26	H36
'1/2 hard'	H14	H24	H34
'1/4 hard'	H12	H22	H32
Annealed	O		
'As-manufactured'	F		

quarter, half, three-quarter and fully-hard. The official system specifies the temper by means of an H-number, such as H14, for example.

Table 4.2 lists the H-numbers in common use. The second digit after the H is the important one, since this defines the actual hardness. Usually this is an even number, the corresponding hardness being obtained by dividing by 8 (e.g. H16 = 6 ÷ 8 = 3/4 hard). For special requirements, material can be supplied to an intermediate level of hardness, in which case the second digit is odd (e.g. H13).

The first digit after the H is of less direct interest, as it merely shows the procedure used by the manufacturer to bring the material to its final hardness. Possible procedures are:

- *Temper-rolled (H1x)*. The material is cold reduced to the required hardness, with no subsequent heating.
- *Temper-annealed (H2x)*. The material is over-worked in the final pass, and is then brought back to the required hardness by a partial anneal.
- *Stabilized (H3x)*. The material is cold reduced to the right hardness, and its properties are then stabilized by a relatively low temperature application of heat. Such treatment may be necessary to prevent the age-softening to which some work-hardened alloys are prone.

4.1.4 The O and F conditions

There are two other possible conditions in which material can be supplied, defined as follows: O, annealed; F, 'as-manufactured'. The O condition defines material that has been fully annealed (i.e. softened) by heating. The stronger F condition applies to products that are hot-formed to their final shape, without any subsequent cold-work or heat-treatment. It typically applies to hot rolled plate, or to sections in the as-extruded condition.

The properties of F condition material cannot be controlled to the degree that is possible with other conditions, and material specifications only quote 'typical' strength values for such material, rather than guaranteed minima. In the case of extrusions, the F condition strength will be only slightly above that for O. With plate, it is often considerably higher, but not reliably so.

Often material ostensibly in the O condition gets very slightly cold worked after annealing, as might occur in a straightening or flattening operation. Such material is officially referred to as being in the H111, rather than the O condition. However, the quoted properties are not usually any different, and the use of the designation O for all such material is unlikely to cause confusion.

4.1.5 Relation between temper and tensile strength

For work-hardened sheet material, there is a fairly consistent relation between temper designation (H-number) and minimum tensile strength (f_u), as laid down in BSEN.515.

The fully-hard temper (H18, H28 or H38), referred to as Hx8, is described as the 'hardest temper normally produced', and for any given alloy it is defined in terms of a specific amount by which f_u exceeds the value f_{uo} for the same alloy in the annealed (O) condition. The intermediate tempers are then defined by taking equal steps in f_u between the two extreme conditions (O and Hx8). The resulting value for f_u is effectively given by the following expression (plotted in Figure 4.1):

$$f_u = Pf_{uo} + Q \tag{4.1}$$

in which the coefficients P and Q are a function of the temper (Table 4.3). This expression is based on the BSEN.515 rules and is valid for any composition having an annealed tensile strength greater than about 90 N/mm^2 (i.e. excluding pure aluminium). In practice the actual material standard

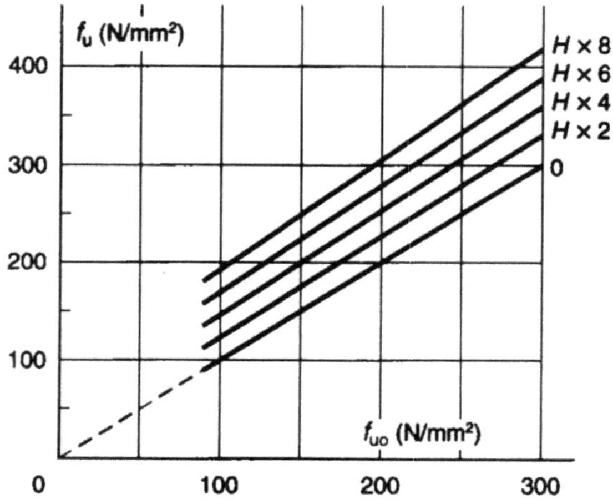

Figure 4.1 Relation between temper and minimum tensile strength (f_u) for work-hardened materials. f_{uo} = minimum tensile strength in the O condition.

Table 4.3 Coefficients in equation (4.1)

Temper	P	Q (N/mm^2)
0	1.00	0
H12, H22, H32	1.03	20
H14, H24, H34	1.06	40
H16, H26, H36	1.09	60
H18, H28, H38	1.12	80

BSEN.485 for flat material lists values of f_u that agree well, but not precisely, with those predicted in the above manner.

For material of given hardness, the decision whether to use the H1x, H2x or H3x condition is to some extent a matter of what the supplier will provide. Only one of the three may be readily available (i.e. from a stockist). Generally speaking, the H2x and H3x conditions have the same tensile strength as the H1x, but with slightly lower proof stress (f_o) and higher elongation (*el*).

4.1.6 Availability of different tempers

In the new material standard (BSEN.485), non-heat-treatable alloys are generally shown as being manufactured in the full range of possible conditions. For each alloy (almost), strength data are provided for all three types of temper (H1x, H2x and H3x), and also for the full range of hardnesses in each type (Hx2, Hx4, Hx6, Hx8). But just because every conceivable condition is covered in the standard, this does not mean that they are all readily obtainable. For small orders, or with material supplied by stockists, the availability will be much restricted.

The 1xxx and 3xxx-series materials are fairly readily available in the temper-rolled conditions right up to fully-hard (H12 to H18). Some are even offered as 'super-hard' sheet in an H19 temper, which is stronger than H18 (but less ductile).

The temper availability of the 5xxx alloys is more restricted, because the hardest tempers are difficult to produce and can only be offered in large tonnages. The 5xxx-series sheet is normally produced to a maximum hardness of Hx4 (or Hx2 in the case of the 5083 alloy), typically in a temper-annealed condition (H2x). The harder tempers (Hx6, Hx8) have attractive strength figures, comparable with those of 6xxx-series heat-treated, but can only be considered if a large quantity is involved.

4.1.7 Heat-treated material

Wrought products in the heat-treatable alloys are strengthened by heat-treatment after they have reached their final form. This comprises *solution treatment* followed by *ageing*. The solution treatment consists of first heating the metal to a temperature of some 500°C, which causes the alloying constituents to go into solid solution; and then quenching. Immediately after the quench these constituents remain dissolved, but with the passage of time they gradually precipitate out in the form of small hard clusters, which impede the movement of dislocations under stress, thereby raising the strength, a process known as *age hardening*.

Ideally, the solution treatment quench takes place in a tank from a precisely controlled temperature, which depends on the alloy. However, with the 6xxx series, the exact quenching temperature is less critical and for extrusions in these alloys it is usually acceptable simply to *spray quench* the metal as it comes out of the die, thus saving cost. With some 6xxx extrusions, it is even possible to *air quench* with the spray turned off, and still obtain useful properties.

After the solution treatment there are two options. Either the material can be left to age naturally at room temperature over a period of days (*natural ageing*), or it can be heated in an oven at a temperature of about 150–180°C, this being known as *artificial ageing* or *precipitation treatment*. The advantages of the latter are that the material ends up stronger (but less ductile), and the final properties are achieved in hours rather than days.

The condition (or temper) of heat-treated material is specified by means of a T-number. Possible conditions range from T1 to T10, as defined in BSEN.515. Three are of interest in structural design:

- T4 solution treatment, followed by natural ageing;
- T5 air quench, followed by artificial ageing (applicable only to some 6xxx-series extruded material);
- T6 solution treatment, followed by artificial ageing.

T6 is the fully-heat-treated condition (maximum strength). T4 is more ductile and is selected when formability is a factor. T5 may be chosen for very thin extrusions, which would distort excessively if subjected to a water quench.

Recent specifications distinguish between material that is not subjected to a straightening process after quenching, and that which is. For plate or sheet, the latter condition is indicated by adding the digits 51 to the basic T4 or T6, while for extrusions and drawn tube the additional digits are 510. In practice, this is a fairly academic distinction, since the same properties are usually quoted for either condition. Therefore designers will not go far wrong by simply writing T4 or T6.

4.2 CHARACTERISTICS OF THE DIFFERENT ALLOY TYPES

The general characteristics of the seven alloy series are summarized in Table 4.4. Those most used in non-aeronautical construction are the 5xxx and 6xxx series.

4.2.1 Non-heat-treatable alloys

(a) 1xxx-series alloys

The materials in this series comprise commercially pure aluminium in a range of purities. The third and fourth digits in the reference number define the purity, by indicating the minimum percentage of aluminium over and above 99.00. Thus '1050' refers to a material having a purity of at least 99.50%. Possible purities range from 99.00 to 99.99%.

Pure aluminium is weak, with a maximum possible tensile strength of about 150 N/mm^2. It is selected when corrosion resistance is critical, as in chemical plant. The higher the purity the better the corrosion resistance, but the lower the strength. In the O or F condition, pure aluminium is relatively soft, and is used when high formability is needed. The most ductile version is 'super-purity' (99.99% pure aluminium), which is produced from lower purity metal by a further 'zone-refining' process.

(b) 3xxx-series alloys

The relatively low manganese content of these alloys, sometimes with added magnesium, makes them half as strong again as pure aluminium, while retaining a very high resistance to corrosion. Tensile strengths go up to 200 N/mm^2 or more. In construction, the main application (in the fully-hard temper) is for profiled sheeting, as used in the cladding of buildings and other structures. 3xxx-series strip is also employed for making welded tube (irrigation pipe, for example).

(c) 4xxx-series alloys

These Al-Si alloys are omitted from Table 4.4 because they seldom appear in the form of actual members. Their usage is for castings (Section 4.5), and weld filler wire (Section 4.6.2). At one time architectural extrusions were occasionally produced in material of this type, because of the attractive dark finish that can be produced on it by natural anodizing.

(d) 5xxx-series alloys

These represent the major structural use of the non-heat-treatable alloys. The magnesium content varies from 1 to 5%, often with manganese added, providing a range of different strengths and ductilities to suit different applications. Corrosion resistance is usually excellent, although it is possible for the stronger versions to suffer abnormal forms of corrosion when operating in hot environments.

The 5xxx alloys appear mostly as sheet or plate. At the lower end of the

Table 4.4 Characteristics of different alloy series

	Non-heat-treatable			Heat-treatable				
Alloy series Main ingredients	1xxx (Pure)	3xxx Mn	5xxx Mg	2xxx Cu, etc.	6xxx Mg, Si		7xxx Zn, Mg	
Type					Weaker	Stronger	Weaker	Stronger
Top tensile strength (approx.) in N/mm^2	150	200	300	450	200	300	350	550
Durability rating	A	A	A*	D	B	B	C	D
Arc-welding	Yes	Yes	Yes	No	Yes	Yes	Yes	No
Extrudability	Very good	–	Moderate to poor	Poor	Very good	Good	Moderate	Poor
Available forms	All	Sheet, strip	All	All	Extrusions, tube	All	All	All

Note. * The stronger alloys in the 5xxx series can have corrosion problems when operating for a long time in a hot environment.

range, they have good formability and are the natural choice for sheet-metal fabrications. At the upper end, they are employed in welded plate construction, typically in the F-condition. Such plate is tough and ductile, with possible tensile strengths exceeding 300 N/mm^2.

The 5xxx series is little used for extrusions, which are only available in the O and F conditions with a rather low proof stress. Extrudability is poor compared to 6xxx and very thin sections are impossible. Also, the use of bridge-dies is ruled out, so that the only way to extrude a hollow section is over a mandrel (Section 2.3.5). An acceptable practice is to employ 5xxx-series plating with 6xxx extrusions as stiffeners.

4.2.2 Heat-treatable alloys

(a) 2xxx-series alloys

This comprises a range of alloys all containing copper, together with other possible elements such as Mg, Mn and Si. Wilm's original 'duralumin' was an alloy of this type (Section 1.4.4). The 2xxx-series comprises high-strength products, and is largely confined to the aerospace industry. Material for this market is mainly supplied to special Aerospace Standards, with closer control of manufacture than that required by the ordinary 'general engineering' ones, thus pushing up the cost.

In the naturally aged T4 condition, these alloys have mechanical properties similar to mild steel, with a typical proof stress of 250 N/mm^2, ultimate tensile strength approaching 400 N/mm^2 and good ductility. In the full strength T6 condition, the proof and ultimate stress can reach 375 and 450 N/mm^2 respectively, but with reduced ductility. The good mechanical properties of the 2xxx-series alloys are offset by various adverse factors, such as inferior corrosion resistance, poor extrudability, unsuitability for arc welding, and higher cost. However, the corrosion resistance of thin material can be improved by using it in the form of *clad sheet* (Section 2.2.5).

The American civil engineering structures of the 1930s were all in 2xxx-type alloy, using the ductile T4 temper (Section 1.5.3). After 1945, the 2xxx series was superseded by 6xxx for such use, despite the lower strength of the 6xxx series. Today there is little non-aeronautical use of 2xxx alloys.

(b) 6xxx-series alloys

These materials, mainly containing Mg and Si, have the largest tonnage use of the heat-treatable alloys. They combine reasonable strength with good corrosion resistance and excellent extrudability. Adequate solution treatment of extrusions can usually be obtained by spray quenching at the die, and air quenching is possible with some of the alloys (Section 2.3.2). The 6xxx materials are readily welded, but with severe local softening in the heat-affected zone. Broadly they divide into a stronger type and a weaker type.

The stronger type of 6xxx material in the T6 condition is sometimes described as the 'mild steel' of aluminium, because it is the natural choice for stressed members. In fact it is a weaker material than mild steel, with a similar proof or yield stress (250 N/mm^2), but a much lower tensile strength (300 N/mm^2). It is also less ductile.

The weaker type of 6xxx alloy, which is not normally offered as sheet or plate, is the extrusion alloy *par excellence*. It is more suitable than any other alloy for the extrusion of thin difficult sections, and is a common choice for members that operate at a relatively low stress level, especially when good surface finish is important. Typical examples are when design is governed by stiffness (rather than strength) as with many architectural members, or when fatigue is critical, as for example in the structure of railcars.

(c) 7xxx-series alloys

These alloys mainly contain zinc and magnesium. Like the 6xxx-series, they can be either of a stronger or a weaker type. The stronger type embraces the strongest of all the aluminium alloys, with tensile strengths in the T6 condition up to 550 N/mm^2. The application of such alloys is almost entirely in aircraft. They have inferior corrosion resistance, poor extrudability and are unsuitable for arc welding. As with 2xxx alloys, sheet can be supplied in clad form to prevent corrosion (Section 2.2.5).

The weaker type of 7xxx material is a very different proposition and in the non-aeronautical field is a valid alternative to the stronger type of 6xxx material, especially for welded construction. It has superior mechanical properties to 6xxx material, and HAZ softening at welds is less severe. But corrosion resistance, although much better than for the stronger kind of 7xxx alloy, is not as good as with 6xxx alloy. Also there can be a possibility of stress-corrosion. Extrudability is not quite as good as for the 6xxx series, but hollow (bridge-die) extrusions are still possible. An important factor, when using 7xxx-series alloy as against 6xxx alloy, is the need for greater expertise in fabrication.

4.3 DATA ON SELECTED WROUGHT ALLOYS

4.3.1 How mechanical properties are specified

Aluminium material standards quote two levels of stress, both of which must be attained for a batch of material to be accepted:

f_o minimum value of the 0.2% proof stress (or '0.2% offset');
f_u minimum tensile strength (or 'ultimate stress').

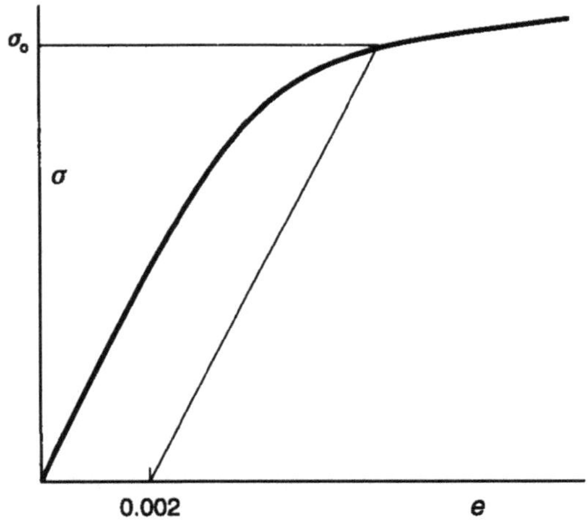

Figure 4.2 Construction for obtaining 0.2% proof stress σ_0 from the stress–strain curve.

These are checked in mill tests, using coupons pulled in tension. The testing machine in effect generates a stress–strain curve, and the proof stress is determined by means of the well-known construction shown in Figure 4.2. The tensile strength is taken as the fracture force divided by the original section area of the coupon. Sometimes the standard fails to give specified values and one has to make do with 'typical' ones, which are not binding on the manufacturer.

The compressive proof stress is never recorded, and a designer would normally assume it to be the same as in tension. This is often a reasonable assumption, but may be incorrect for non-heat-treatable extrusions which are hot-finished and then stretch-straightened. Because of the Bauschinger effect, the tensile proof stress is likely to be enhanced and the compressive proof depressed by so doing, the difference between the two values being as much as 20% in some cases. This need not normally worry a designer, because the f_0-value quoted for such material tends to be conservative.

Standards also call for a minimum percentage elongation *el* that any batch must achieve. This is typically measured on a gauge-length of 50 mm and gives a crude indication of ductility.

4.3.2 Specific alloys and their properties

It is impossible in a book such as this to provide strength data for all the aluminium materials that exist. Not only are there scores of registered compositions in each alloy series, but their properties vary according to form of product, temper and thickness. It is essential for a designer to refer to an appropriate standard to find precise strength values for the materials he/she is planning to use. Relevant standards, covering a total of some

60 alloys, are:

BSEN.485	plate and sheet;
BSEN.755	extruded sections;
BSEN.754	drawn tube.

Figures 4.3 and 4.4 show compositions and minimum mechanical properties for a representative range of aluminium materials, mainly based on the above standards. Where a band is drawn, this shows the possible range for the property concerned, depending on form and thickness.

Table 4.5 presents more specific data for a shortlist of alloys, covering plate, sheet and extrusions. The strength of drawn tube can be estimated by taking that for either equivalent work-hardened sheet material, or else equivalent heat-treated extrusion material.

4.3.3 Comments on certain alloys

(a) 5xxx-series alloys

The A-rating for durability accorded to these alloys (Table 4.4) is usually a true indication of their excellent corrosion resistance. However, when such alloys have a magnesium content higher than about 3.5 or 4%, and they are required to operate in a hot environment, there is a danger of abnormal forms of corrosion (Section 4.7.1). For such applications, it is desirable to keep the magnesium content down by using a weaker alloy. For example, in the construction of welded tipper-truck bodies one would normally select one of the strongest versions of 5xxx-series alloy, such as 5083. But this alloy, with a nominal magnesium content of 4.5%, becomes unacceptable if the body has to carry hot material. In such a case, one should change to a weaker alloy with less magnesium.

The following 5xxx-series alloys, not covered in Table 4.5, also appear in structural codes:

5052 (Mg 2.5, Cr 0.2)	slightly stronger than 5251;
5754 (Mg 3.1)	intermediate between 5251 and 5154A;
5454 (Mg 2.7, Mn 0.7, Cr 0.1)	similar strength to 5154A.

(b) 6xxx-series alloys

As already explained, these alloys are broadly available in a stronger and a weaker version. The latter with its superb extrudability is selected when stiffness is the main criterion, the 6063 alloy being much used. For more highly stressed situations, the stronger version becomes necessary, popular alloys being 6082 in Britain, and 6061 in North America. The 6061 is not quite as strong as 6082, but is slightly better for forming. Other

Table 4.5 Minimum strength data for a shortlist of alloys

Alloy/condition		Plate, sheet			Extrusions			
		t mm	f_o MPa	f_u MPa	t mm	f_o MPa	f_u MPa	*el* %
3103	0	<50	35	90	–	–	–	17–28
Mn 1.2	H14	<25	120	140	–	–	–	2–5
	H18	<3	165	185	–	–	–	1–2
5251	0	<50	60	160	all	60	160	13–18
Mg 2.0, Mn 0.3	H22	<25	120	190	–	–	–	4–12
	H24	<12	140	210	–	–	–	3–10
	H26	<4	170	230	–	–	–	3–7
5154A	0	<50	85	215	<25	85	200	12–16
Mg 3.5	H22	<40	180	250	–	–	–	5–9
	H24	<25	200	270	–	–	–	4–7
	H26	<6	230	290	–	–	–	3–5
5083	0	<50	125	275	all	110	270	10–16
Mg 4.5, Mn 0.7	H22	<40	215	305	–	–	–	5–10
Cr 0.15	H24	<25	250	340	–	–	–	4–7
6063	T4	–	–	–	<25	65	130	12–14
Mg 0.7, Si 0.4	T5	–	–	–	<3	130	175	5–8
	T5	–	–	–	3–25	110	160	6–8
	T6	–	–	–	<10	170	215	6–8
	T6	–	–	–	10–25	160	195	6–8
6061	T4	<12	110	205	<25	110	180	12–18
Mg 1.0, Si 0.6	T6	<12	240	290	<25	240	260	6–10
Cu 0.25, Cr 0.2								
6082	T4	<12	110	205	<25	110	205	12–14
Mg 0.9, Si 1.0	T5	–	–	–	<5	230	270	6–8
Mn 0.7	T6	<6	260	310	<5	250	290	6–10
	T6	6–12	255	300	5–25	260	310	6–10
7020	T4	<12	210	320	–	–	–	11–14
Zn 4.5, Mg 1.2	T6	<12	280	350	40	290	350	7–10
Cr 0.2								
7019	T6	–	330	380	–	330	380	8
Zn 4.0, Mg 2.0								

Note: MPa is the same as N/mm^2

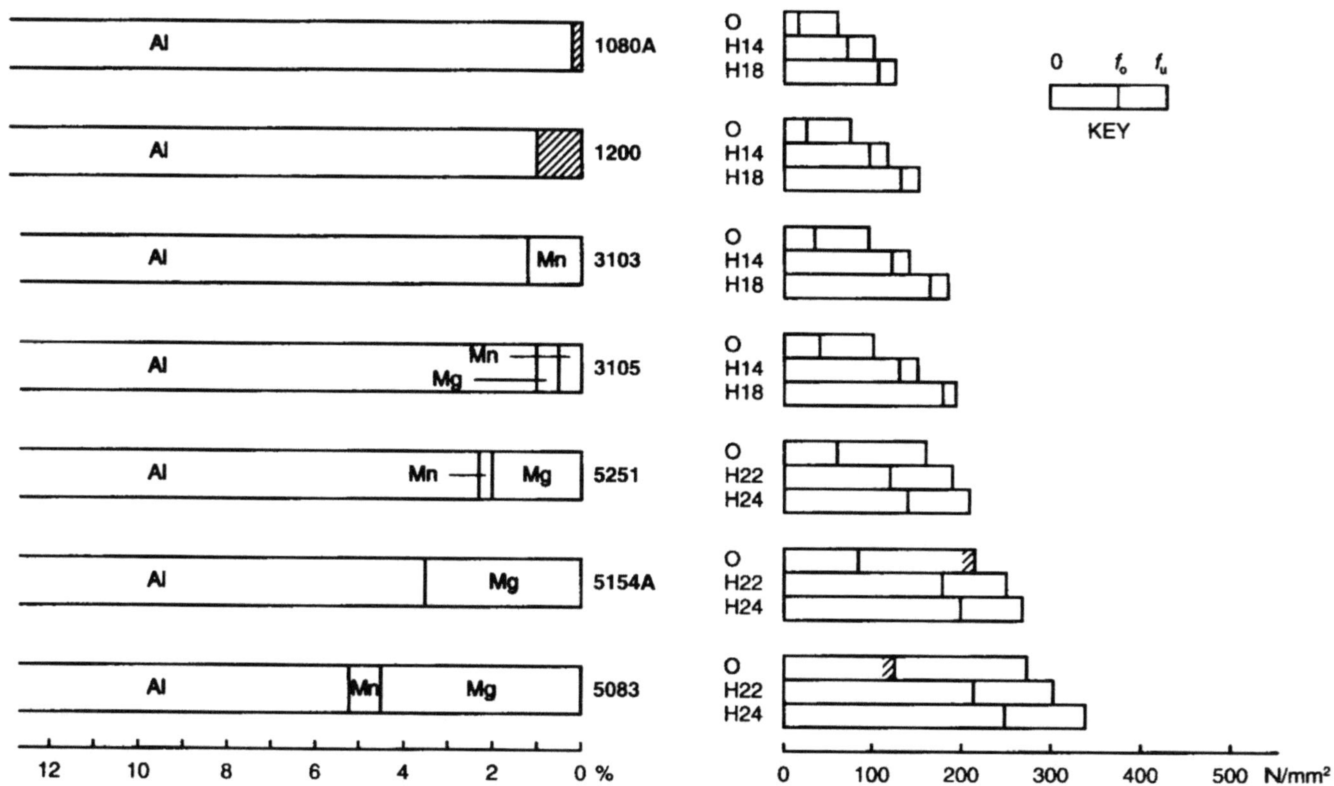

Figure 4.3 Nominal composition and minimum properties (f_o, f_u) for a range of non-heat-treatable materials.

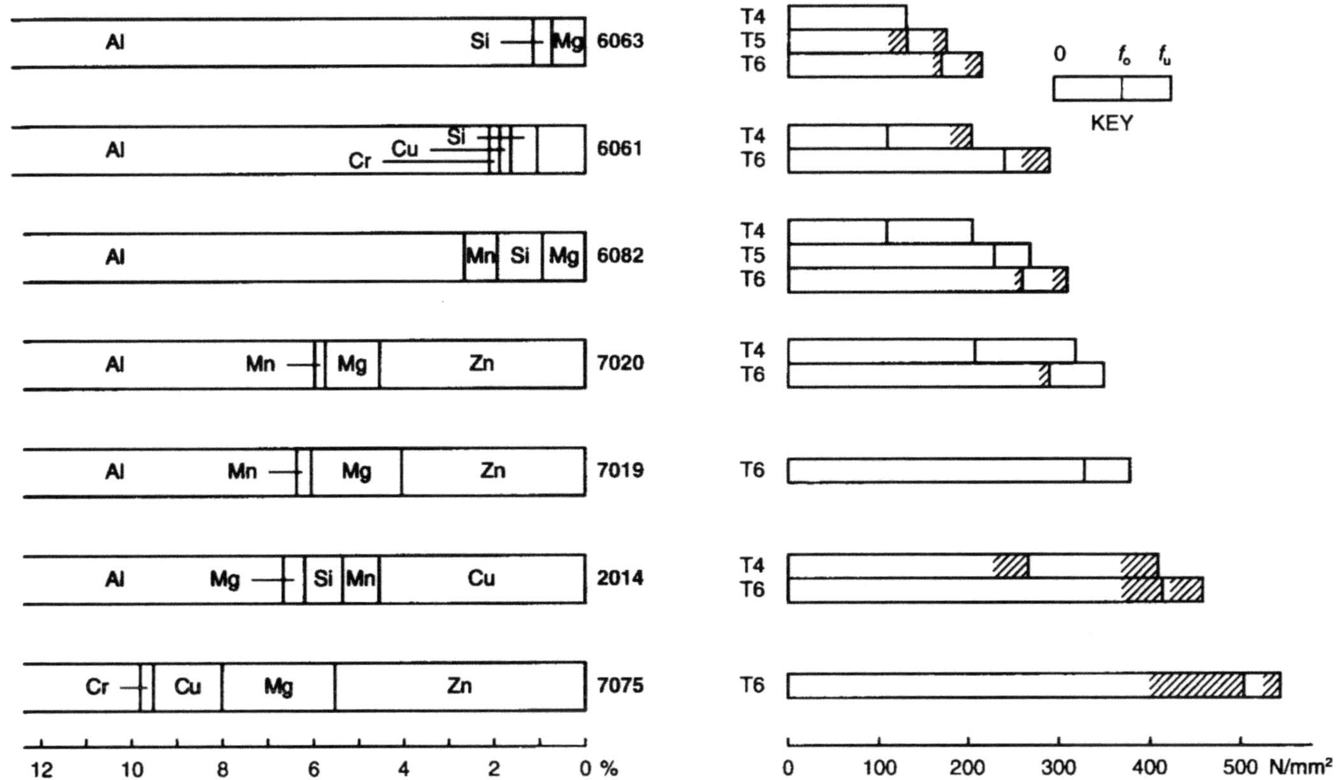

Figure 4.4 Nominal composition and minimum properties (f_o, f_u) for a range of heat-treatable materials.

popular alloys in the series are:

6060 (Mg 0.5, Si 0.4, Fe 0.2)	slightly weaker than 6063, and even more extrudable;
6005A (Mg 0.5, Si 0.7)	a compromise alloy, lying between the stronger and weaker kinds of 6xxx material. Somewhat weaker than 6061.

(c) Weldable 7xxx-series alloys

The composition of the weldable (weaker) 7xxx-series alloys is adjusted to obtain a compromise between strength and weld integrity. If such material is over-alloyed, stress corrosion and cracking are likely to occur at the welds. The alloy 7020 has been carefully designed to give good strength with reasonable freedom from cracking, and an alloy such as this would be the one normally recommended. The use of a stronger version, in order to save weight, can only be considered if suitable precautions are taken during fabrication, following specialist metallurgical advice. The 7019 alloy, which probably represents the top end of the practicable strength range for welded construction, has been successfully used in large tonnages for military bridging.

4.3.4 Minimum bend radius

The bendability of flat aluminium material varies greatly from alloy to alloy. The problem in fabrication is to know the minimum bend radius (r), this being a function of the alloy, temper and metal thickness (t). Material standards list minimum values of r/t, but these are only 'recommended' and not binding on the supplier. Here we present a rationalized procedure for obtaining r/t, broadly based on the data given in BSEN.485. It must be

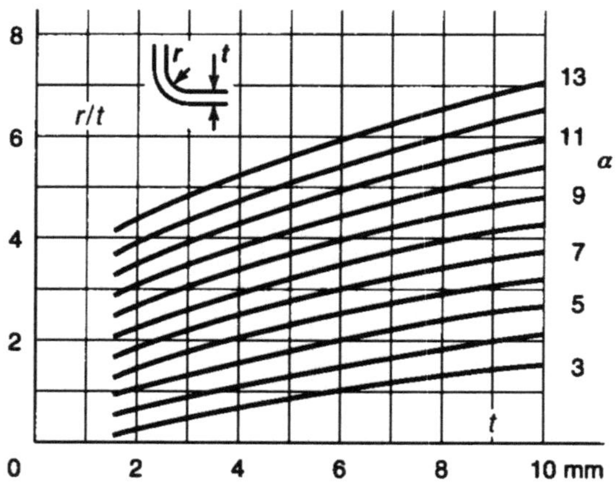

Figure 4.5 Approximate variation of minimum r/t with t for 90° bends in flat material. Values of the bendability index a are listed in Table 4.6.

Table 4.6 Bendability index α for flat material

Alloy	Temper	α	Temper	α	Temper	α
1200	0	3	H12	5	H14	6
3103	0	4	H12	5	H14	6
5251	0	5	H22	6	H24	7
5154A	0	6	H22	8	H24	9
5083	0	6	H22	8	H24	10
6061	T4	8	T6	11		
6082	T4	9	T6	13		

appreciated that in terms of practical fabrication the subject of bend radii is not an exact science.

For any given thickness the minimum 90° bend radius may be estimated with the aid of Figure 4.5, which gives curves of r/t plotted against t. The appropriate curve depends on the bendability index α of the material being bent, values thereof being listed in Table 4.6 for a range of alloys. The equation to the curves in the figure is:

$$\frac{r}{t} = A + B\sqrt{t} \tag{4.2}$$

where: $A = 0.3\alpha - 1.7$; $B = 0.08\alpha + 0.5$; and t = thickness in mm.

The curves in Figure 4.5 are valid for cold bending in thicknesses from 1.5 to 10 mm. They apply to 90° bends. For bends beyond 90°, the minimum radius will increase, especially in the harder tempers. With work-hardened material, the bendability is slightly better for the H2x and H3x tempers, than it is for H1x. For heat-treated material in the T4 temper, we have assumed that forming takes place after the material has artificially aged to its final condition; a smaller r/t will be possible if it is bent immediately after quenching.

The bending of extruded sections is an altogether more complicated subject, and the reader is referred to the useful publication *Aluminium Extrusions – A Technical Design Guide*, issued by the Shapemakers. (Section 1.6)

4.3.5 Strength variation with temperature

One weakness of aluminium as a structural metal is its fairly rapid fall-off in strength with temperature. On the other hand, it has the advantage that with decreasing temperature the properties steadily improve, and there are none of the brittle fracture problems met with in steel. The tensile strength of 6082-T6 goes down by 70% at 200°C, compared to room temperature, but increases by 40% at −200°C.

Figure 4.6 shows how tensile strength varies with temperature (T) for a range of alloys. The curves are plotted from data supplied by the Aluminium Federation, based on specimens held for a long time at the

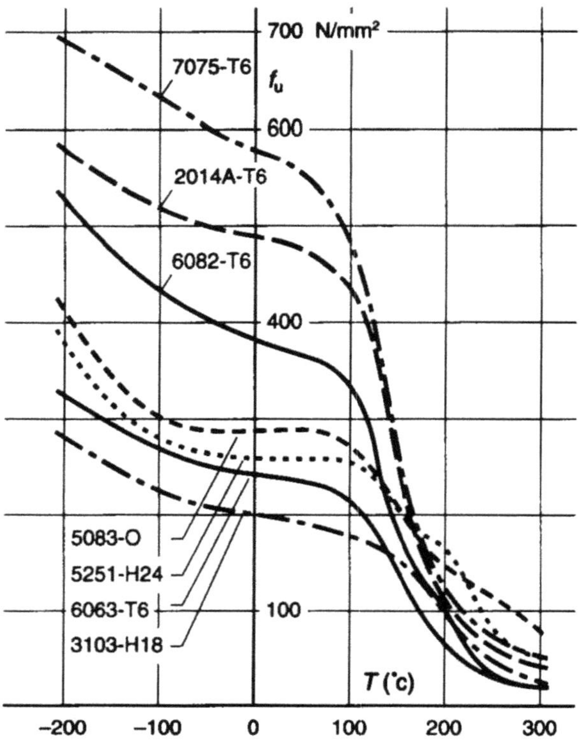

Figure 4.6 Variation of tensile stress (f_u) with temperature. Tested at temperature T after long-term exposure at that temperature.

relevant temperature and then tested at that temperature [1]. The reason for the room-temperature values not agreeing exactly with those given previously is that they were obtained from typical material, whereas the Section 4.3.2 data were based on specification minima. Figure 4.7 gives more comprehensive data for two of the alloys [13], from which we see that the loss of strength at temperatures over 100°C is considerably more pronounced with 6082 than it is with 5083. Weakening with temperature is a significant factor in the design of supersonic aircraft, and special high-strength alloys have been developed to reduce the effect. One such alloy is 2618, which under prolonged heating at 200°C has twice the strength of 2014A (in the T6 temper).

Also of interest is the decrease in strength after a relatively short-term exposure to elevated temperature (Figure 4.8). This is relevant to the fire-rating of buildings and other structures.

4.3.6 Properties of forgings

Aluminium forgings range in size from small components, such as the hinges, etc., on truck bodies, to the main members forming the

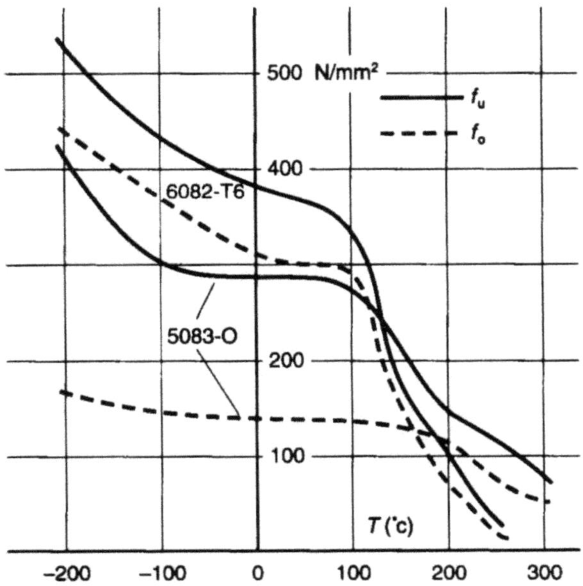

Figure 4.7 Variation of proof stress (f_o) and tensile strength (f_u) with temperature, for two alloys. Procedure as for Figure 4.6.

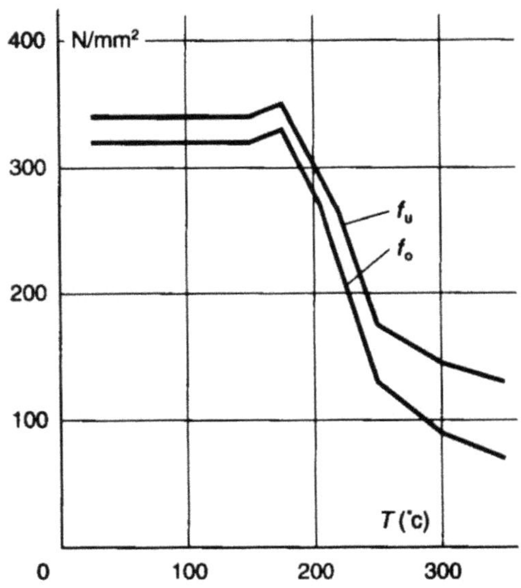

Figure 4.8 Variation of room-temperature properties after two-hour exposure to an elevated temperature T for a typical batch of 6082-T6 material. f_o = 0.2% proof stress, f_u = tensile strength. (Alcan.)

undercarriages of aircraft. They are specialist products on which it is essential to confer with the supplier.

Forgings are available in many of the usual wrought alloys. Non-heat-treatable ones are supplied in the 'as-manufactured' F condition in materials such as 5083, 5454 and 5251. The typical strength of these, which will vary across the component, tends to be higher than the O condition values for plate or extruded material. Heat-treated forgings can be obtained in alloys such as 6082 and 2014, with properties in the T4 and T6 conditions comparable to those for other wrought products in those alloys.

4.4 STRESS–STRAIN CURVES

4.4.1 Empirical stress–strain relation

The shape of the stress–strain (σ, e) curve is not mentioned in material specifications, nor does it directly figure in design. But it is useful for a designer to have a feel for it. Sometimes specialist computer programs are employed, either for structural analysis or in research studies. Such programs require the use of a suitable mathematical expression to relate stress and strain, a convenient empirical relation being the well known Ramberg–Osgood equation:

$$e = \frac{\sigma}{E} + 0.002\left(\frac{\sigma}{\sigma_0}\right)^n \tag{4.3}$$

where E = modulus of elasticity, and σ_0 = actual 0.2% proof stress.

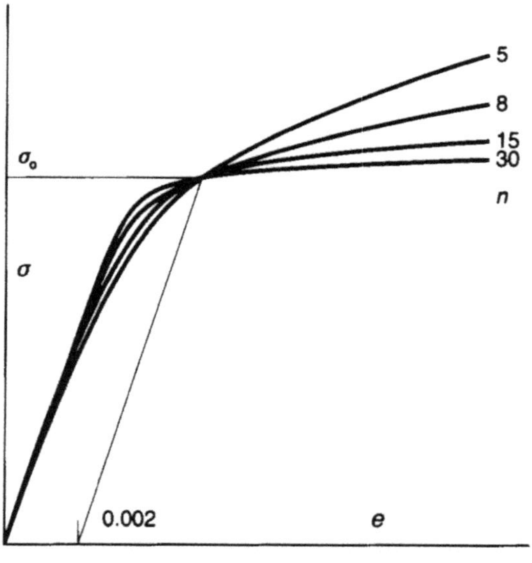

Figure 4.9 Ramberg–Osgood stress–strain equation (4.3), effect of n.

The two terms in this expression are respectively the elastic and plastic strains. It will be seen (Figure 4.9) that the curve produced automatically satisfies two necessary requirements, whatever the value of n:

1. There is (effectively) a linear elastic range.
2. The plastic strain correctly equals 0.002 when the stress reaches σ_o.

The role of the index n is to control the curvature of the knee on the curve. A high n produces a curve with a relatively abrupt knee, a low n one with a more rounded knee. One can thus refer to 'high-n' and 'low-n' materials. n typically ranges from 40 down to 5 for aluminium materials.

4.4.2 Stress–strain curve for minimum strength material

For any given alloy the actual properties of the material (as supplied) will vary considerably from batch to batch, and also along the length of an extrusion. For analytical purposes, it may be necessary to define a stress–strain curve that would represent the response of *minimum strength* material, i.e. material that just attains the specification values for f_o, f_u and *el*. Referring to Figure 4.10, a curve is needed that goes through the proof-stress point A and also the fracture point B. This is achieved by putting $\sigma_o = f_o$ in equation (4.3), and taking n as follows:

$$n = \frac{\log(5el)}{\log(f_u/f_o)} \tag{4.4}$$

The curvature of the knee, defined by n, is strongly dependent on the ultimate-proof ratio f_u/f_o but is insensitive to the elongation *el*. A material with a low value of f_u/f_o will exhibit a relatively abrupt knee (high n), whereas for material having a high f_u/f_o the knee will be more rounded

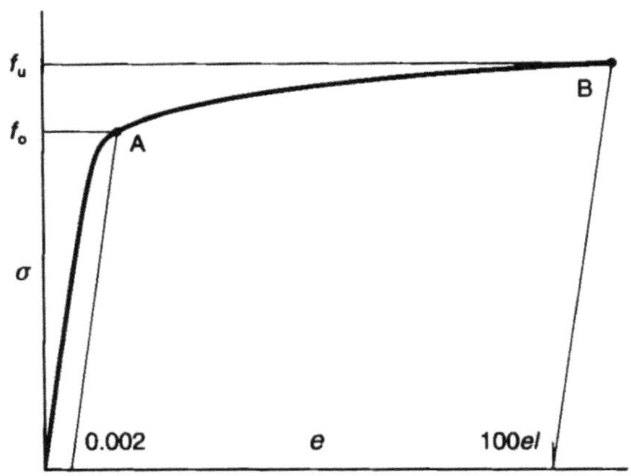

Figure 4.10 Construction for obtaining a stress–strain curve to represent minimum strength material.

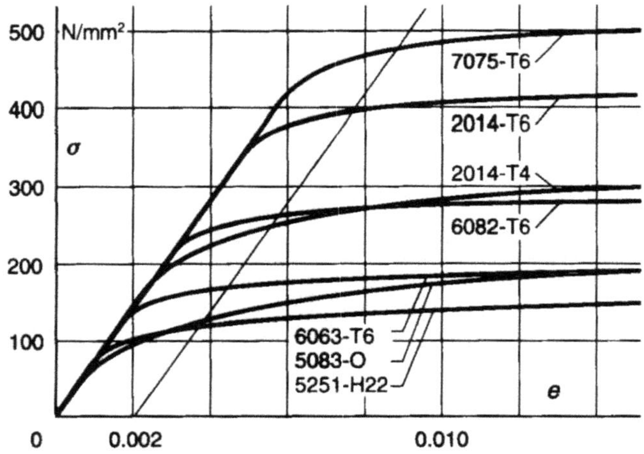

Figure 4.11 Minimum stress–strain curves for a selection of aluminium materials, based on a thickness of 6 mm, generally for flat material. The 6063 curve is for 6 mm thick extruded profile.

(low n). Much of aluminium construction is in materials that are either fully heat-treated, or else in a heavily work-hardened condition. These typically have a low ultimate/proof ratio (below 1.2) and hence a stress–strain curve with a fairly abrupt knee. A higher value for f_u/f_o, corresponding to a more rounded curve, would be exhibited by material in the T4 or annealed condition.

Figure 4.11 gives curves constructed by the above method for a selection of aluminium materials.

Mazzolani discusses the use of more precise mathematical expressions for representing the stress–strain relation [26].

4.5 CASTING ALLOYS

Compared to forgings, castings are generally less ductile and weaker, although tensile strengths up to 300 N/mm^2 are possible. They are cheaper than forgings. The design of aluminium castings is beyond the scope of this book, but below we give some limited data on certain alloys.

4.5.1 Numbering system

The unified system for specifying cast material, now adopted in Europe, is described in BSEN.1706. For a given material it uses a two-part reference, such as AC.51400-F or AC.42000-T6, in which the first part refers to the alloy, and the second part to the condition. In Britain, the new European alloy numbers for castings supersede the old 'LM' system, although the actual compositions are broadly the same.

The first digit in the number indicates the alloy series, which depends on the main alloying ingredients in the same way as for wrought alloys (Table 4.1). Thus the AC.4xxxx series covers alloys containing just Si, and the AC.5xxxx those with substantial Mg. Unlike wrought, the European numbers for casting alloys do not tally with the American ones.

The symbol or symbols after the hyphen have the same basic meanings as for wrought material:

F 'as-manufactured', i.e. as cast;
T4 solution treated after casting and naturally aged;
T6 solution treated and artificially aged.

4.5.2 Three useful casting alloys

A wide range of casting alloys exists, generally different from those used for wrought products. Table 4.7 provides data for three selected alloys, which should form part of a designer's repertoire. The values quoted in the table are minimum strengths to be achieved by separately cast tensile specimens. They do not accurately represent the metal in the actual casting, the properties of which are unlikely to be uniform. Alternative strength values are given, depending on whether a sand mould is used or a permanent one (chill cast), the chill-cast strength being always higher.

AC.44100

This is a non-heat-treatable alloy, available in the as-cast F condition. It is the most popular casting alloy in aluminium, because of its exceptional fluidity. It is suitable for intricate shapes and can cope with thicknesses down to about 2.5 mm. At the other end of the scale, it can be used for very large castings, up to 3000 kg in sand and 200 kg in permanent mould. It has good corrosion resistance and is successfully employed in 'on-deck' marine applications.

Table 4.7 Three selected casting alloys

BSEN number	BS 1490	Nominal composition (%)	Durability rating	Condition	Minimum f_u (N/mm^2)	Elongation (%)
AC.44100	LM6	Si 11.5	B	F sand	160	5
				F chill	190	7
AC.51400	LM5	Mn0.5, Mg4.5	A	F sand	140	3
				F chill	170	5
AC.42000	LM25	Si7.0, Mg0.4	B	F sand	130	2
				F chill	160	3
				T6 sand	230	2
				T6 chill	280	2

AC.51400

This is another non-heat-treatable alloy. It finds application instead of AC.44100 when appearance is critical, because of its exceptional corrosion resistance and suitability for anodizing. But, against this, it has inferior fluidity, ruling out its use for thin intricate castings. Good foundry practice is important.

AC.42000

This heat-treatable alloy, in the T6 condition, is chosen when higher strength is needed. Like AC.44100, it has excellent fluidity and good corrosion resistance. Also it is more readily machined.

4.6 ALLOYS USED IN JOINTS

4.6.1 Fastener materials

Chapter 11 includes design data on four typical alloys for use as fasteners, namely 6082, 6061, 5154A and 5056A. The first three have already been mentioned above (Section 4.3). The fourth is an alloy containing a nominal Mg 5.0 and Mn 0.35% that produces fasteners as strong as 6082. Because of its high magnesium content it is unsuitable for prolonged use in hot environments.

Stronger alloys are employed for the fasteners used in aeronautical construction. These can be in 2xxx-series alloy (T4 or T6 temper), or in a strong 7xxx-series material such as 7075-T6. In the USA strong fasteners in these materials also find use outside the aero field.

4.6.2 Weld filler wire

Alloys for use as welding wire are divided into four types (Section 3.3.4). Table 4.8 lists, with their nominal compositions, the alloys of each type that are covered in BS.8118.

4.7 CORROSION

4.7.1 Corrosion of exposed surfaces

Any aluminium surface exposed to air develops a thin oxide film, which is hard, chemically stable and tightly keyed to the metal. Though very thin (typical thickness 0.005 mm), this layer prevents further oxidation. When damaged, it immediately reforms, provided oxygen is available, and it is this that gives aluminium its good durability.

Table 4.8 Nominal composition of weld filler alloys

	Alloy	Si %	Mn %	Mg %	Other	Durability rating
Type 1	1080A	99.8% Al minimum				A
	1050A	99.5% Al minimum				
Type 3	3103	–	1.2	–	–	A
Type 4	4043A	5.2	–	–	–	B
	4047A	12.0	–	–	–	C
Type 5	5356	–	0.12	5.0	+Cr	A*
	5056A	–	0.35	5.0	–	A*
	5183	–	0.75	4.7	+Cr	A*
	5556A	–	0.80	5.2	+Cr	A*
	5154A	–	–	3.5	–	A
	5554	–	0.75	2.7	+Cr	A

Note. * The use of these fillers may lead to corrosion problems under prolonged exposure in a hot environment.

Atmospheric corrosion of aluminium is by a process of *pitting*. Often referred to as 'weathering', this is radically different from the rusting of steel. Air-borne pollutants, such as sulphuric acid and sodium chloride, cause local pits to form, the depth of which may be used as a measure of the severity of the attack. In the first year, the pit depth increases relatively quickly, but with the passage of time the corrosion products effectively stifle the process. Eventually, after five years or so, further corrosion almost stops. This pattern of behaviour is illustrated in Figure 4.12, which shows the likely variation of *average* pit depth with time for 3103 sheet in an outdoor industrial location [32].

The severity of pitting depends on the alloy and the environment. For a very durable alloy, such as 3103, the *worst* pits may have a depth after five years of roughly 0.04 mm in a rural location, 0.08 mm in an inland industrial one and something like 0.16 mm at an industrial site by the sea.

The surface of weathered aluminium always looks much worse than it really is, and the immense volume of the corrosion products (some 200

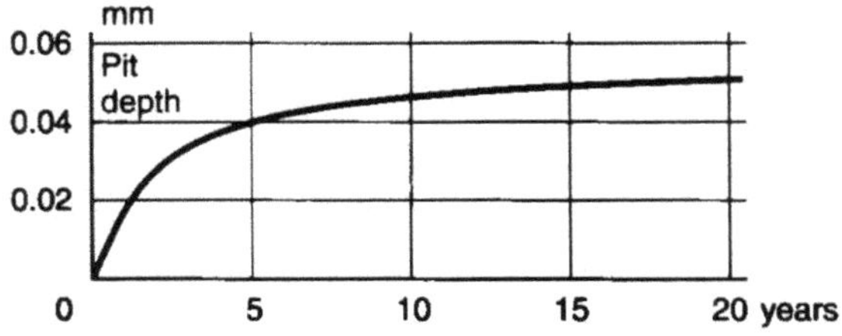

Figure 4.12 Pitting corrosion. Typical growth of average pit depth with time for exposed 3103 sheet in an inland industrial location (Alcan.)

times that of the actual pits) produces a false impression as to the true condition of the metal. Tensile tests on 6063 alloy glazing bar material, after long service at polluted sites, have shown the metal to be only very slightly weakened by the pitting and still well able to meet the specified strength. The only possible worry is the fall-off in ductility that can occur with thin material.

The colour of weathered aluminium varies from a soft blue-grey in rural environments to dark grey or black in industrial ones. In terms of appearance, it is a help if the surface is washed from time to time. When a component has its upper and lower surfaces both exposed to a polluted atmosphere, the lower surface appears to be more heavily attacked, because the corrosion products are not washed away by the rain in the way they are on the upper surface. Regular washing is especially important for an anodized surface, if it is to retain its good looks. Anodizing confers no permanent benefit in appearance unless properly maintained.

Apart from pitting attack, it is sometimes possible for exposed aluminium surfaces to suffer from other more abnormal forms of corrosion, such as: (a) *exfoliation*, namely a delamination or peeling away of layers of metal parallel to the surface, in much the same way as the rusting of steel; and (b) *intercrystalline attack* or 'stress-corrosion'.

Exfoliation is liable to occur with the strong aircraft-type alloys (2xxx, 7xxx), when used unprotected in some adverse environments. Stress-corrosion is a potential hazard with the stronger 5xxx materials (Mg > 3.5%), when they have to operate for a long time at an elevated temperature (say, over 70°C). It is also a consideration in the HAZ at welds made on the weaker type of 7xxx material, such as 7020.

Another hazard is the *poultice effect*, which arises when an aluminium surface is denied oxygen, thus causing it to go on corroding because the oxide layer cannot reform. An example would be when aluminium sheet lies in close contact with certain insulation boards and condensation is present, causing chemicals to leach out of the board and attack the aluminium.

4.7.2 When to protect against corrosion

In most aluminium installations, no protection against surface corrosion is necessary, except for the sake of appearance. When called for, protection can be in the form of painting (Section 3.7.4) or anodizing (Section 3.7.3). Table 4.9 gives recommendations as to when protection is needed in outdoor situations, depending on the durability rating (A, B, C or D) of the alloy and the severity of the environment [9]. The data on ratings A, B and C are taken from BS.8118.

In welded construction, it is possible for the weld filler material, and hence the weld deposits, to be less durable than the parent metal. This must be taken into account when deciding whether or not to paint.

Table 4.9 Pitting corrosion – whether or not protection is needed in outdoor locations

Durability rating	A	B	C	D
Alloy series	1xxx 3xxx 5xxx	6xxx	7xxx(w)	2xxx 7xxx(s)
Inland:				
Rural	No	No	No	Yes
Moderate industrial	No	(Yes)*	Yes	Yes
Severe industrial	Yes	Yes	Yes	Yes
By the sea:				
Rural	No	(Yes)*	Yes	Yes
Moderate industrial	No	(Yes)*	Yes	Yes
Severe industrial	Yes	Yes	Yes	Yes

Notes. 1. In starred cases (*) protection is only needed if metal thickness <3 mm.
2. 7xxx(w) covers the weaker (weldable) kind of 7xxx-series material such as 7020. 7xxx(s) covers the strong kind such as 7075.

4.7.3 Bimetallic corrosion

Table 4.10 shows the order in which common metals lie in the *electrochemical series*, metals near the top being referred to as 'noble' and those at the bottom as 'base'. When two different metals are electrically connected and immersed in an electrolyte, such as water, an electric cell is formed, with ions moving from the baser metal (the anode) to the more noble one (the cathode). Loss of metal occurs at the anode, known as *bimetallic corrosion* (also referred to as 'galvanic' or 'electrolytic' corrosion).

Unfortunately, aluminium is a base metal, superior only to zinc and magnesium in the table. Therefore when aluminium is in contact with most other metals and moisture is present, accelerated corrosion is likely to occur. The only solution is to insulate the two metals from each other.

Table 4.10 Electrochemical series of metallic elements

Noble	Platinum	Pt
	Gold	Au
	Titanium	Ti
	Silver	Ag
	Nickel	Ni
	Copper	Cu
	Tin	Sn
	Lead	Pb
	Iron	Fe
	Cadmium	Cd
	Aluminium	**Al**
	Zinc	Zn
Base	Magnesium	Mg

The severity of bimetallic corrosion of aluminium generally depends on the second metal's position in the series, although there are exceptions. In contact with steel, a common juxtaposition, aluminium will suffer serious corrosion, especially if the surface area of the steel is comparable to that of the aluminium; likewise with cast iron. Suitable precautions are essential (Section 3.7.5). With copper and copper-based metals (brass, bronze), the attack is more severe, and preventative measures are even more vital. The serious effect that copper can have on aluminium is illustrated by the case of a cycle shed once installed at Cambridge University, on which the aluminium sheeting was perforated in a few months by drips of rainwater from a copper roof five storeys above.

With lead, surprisingly, the effect is much less pronounced and preventative measures are only needed in severely polluted environments (industrial or marine). Austenitic stainless steels can safely be used with aluminium, generally without any precautions. With zinc, which lies below aluminium in the table, the zinc rather than the aluminium gets attacked, as with galvanized coatings. The same applies with magnesium.

Bimetallic corrosion of aluminium can sometimes be put to good use in providing *cathodic protection*. Thus, in the North Sea, it is common practice to fix long aluminium bars (anodes) to the steel tubulars of jacket structures in the splash zone. The rate of rusting on the steel is thereby reduced, at the expense of the sacrificial aluminium anodes.

CHAPTER 5

Limit state design and limiting stresses

British Standard BS.8118 follows steel practice in employing the *limit state* approach to structural design, in place of the former *elastic* ('allowable stress') method [14]. Limit state design is now accepted practice in most countries, the notable exception being the USA. In this chapter, we start by explaining the BS.8118 use of the limit state method, and then go on to show how the required *limiting stresses* are obtained.

5.1 LIMIT STATE DESIGN

5.1.1 General description

In checking whether a component (i.e. a member or a joint) is structurally acceptable there are three possible limit states to consider:

- Limit state of static strength;
- Serviceability limit state;
- Limit state of fatigue.

Static strength is usually the governing requirement and must always be checked. Serviceability (elastic deflection) tends to be important in beam designs; the low modulus (E) of aluminium causes it to be more of a factor than in steel. Fatigue, which must be considered for all cases of repeated loading, is also more critical than for steel.

In the USA, when limit state design is mentioned, it is given the (logical) title 'Load and Resistance Factor Design' (LRFD).

5.1.2 Definitions

Some confusion exists because different codes employ different names for the various quantities that arise in limit state design. Here we consistently use the terminology adopted in BS.8118, as below.

- *Nominal loading*. Nominal loads are the same as 'working loads'. They are those which a structure may be reasonably expected to carry in normal service, and can comprise:
  dead loads (self-weight of structure and permanently attached items);
  imposed loads (other than wind);
  wind loads;
  forces due to thermal expansion and contraction;
  forces due to dynamic effects.
  It is beyond the scope of this book to provide specific data on loading. Realistic imposed loads may be found from particular codes covering buildings, bridges, cranes, etc. Wind is well covered. Often the designer must decide on a reasonable level of loading, in consultation with the client.
- *Factored loading* is the factored (up) loading on the structure. It is obtained by multiplying each of the individual nominal loads by a partial factor γ_f known as the *loading factor*. Different values of γ_f can be taken for different classes of nominal load. See Table 5.1.
- *Action-effect*. By this is meant the force or couple that a member or joint has to carry, as a result of a specific pattern of loading applied to the structure. Possible kinds of action-effect in a member are axial tension or compression, shear force, bending moment, and torque. In a joint, the possible action-effects are the force and/or couple that has to be transmitted.
- *Calculated resistance* denotes the ability of a member or joint to resist a specific kind of action-effect, and is the predicted magnitude thereof needed to cause static failure of the component. It may be found by means of rules and formulae given in codes or textbooks, in applying which it is normal to assume minimum specified tensile properties for the material and nominal dimensions for the cross-section. Suggested rules are given in Chapters 8, 9 and 11, based largely on BS.8118. Alternatively the calculated resistance may be determined by testing, in which case the word 'calculated' is something of a misnomer. Testing procedure is well covered in BS.8118.
- *Factored resistance* is the factored (down) resistance of a member or joint. It is the calculated resistance divided by a partial factor γ_m known as the *material factor* (given in Table 5.1).

In discussing limit state design, we use the following abbreviations to indicate quantities defined above:

NA = action-effect arising under nominal loading;
FA = action-effect arising under factored loading;
CR = calculated resistance;
FR = factored resistance ($= CR/\gamma_m$).

In Chapters 8, 9 and 11 the suffix *c* is used to indicate calculated resistance.

Table 5.1 Suggested γ-values for checking the limit state of static strength

Loading factor γ_f		Material factor γ_m	
Sub-factor γ_{f1}		Members	$\gamma_m = 1.2$
Dead load, direct effect	$\gamma_{f1} = 1.2$		
Dead load, countering uplift	0.8	Connections:	
Imposed load, excluding wind	1.33	Mechanical joints	$\gamma_m = 1.2$
Wind load	1.2	Welded joints	1.3–1.6
Force due to temperature effect	1.0	Bonded joints	1.6 min
Estimated force under impact	1.33		
Sub-factor γ_{f2}			
Dead load	$\gamma_{f2} = 1.0$	**Serviceability factor γ_s**	
Imposed, wind or impact load, producing an action-effect on the component that is:		HSFG bolted joints	$\gamma_s = 1.1–1.2$
(i) the most severe	$\gamma_{f2} = 1.0$		
(ii) second most severe	0.8		
(iii) third most severe	0.6		
(iv) fourth most severe	0.4		
Force due to temperature	1.0		

Notes. 1. The loading factor is found thus: $\gamma_f = \gamma_{f1}\gamma_{f2}$.
2. The values given for the material factor assume a high standard of workmanship. For welded and bonded joints, the minimum value should only be used when the fabrication meets the requirements of BS.8118: Part 2, or equivalent.

5.1.3 Limit state of static strength

The reason for checking this limit state is to ensure that the structure has adequate strength, i.e. it is able to resist a reasonable static overload, over and above the specified nominal loading, before catastrophic failure occurs in any of its components (members, joints). The check consists of calculating FR and FA for any critical component and ensuring that:

$$FR \geqslant FA \tag{5.1}$$

In order to obtain FR and FA it is necessary to specify values for the partial factors γ_f and γ_m. These are for the designer to decide, probably in consultation with the client. The BS.8118 recommendations are as follows:

1. *Loading factor* (γ_f). This factor, which takes account of the unpredictability of different kinds of load, is taken as the product of two sub-factors as follows:

$$\gamma_f = \gamma_{f1}\,\gamma_{f2} \tag{5.2}$$

γ_{f1} depends on the kind of load being considered, while γ_{f2} is a factor that allows some relaxation when a combination of imposed loads acts on the structure. Table 5.1 gives suggested values for γ_{f1} and γ_{f2}, based on BS.8118. For initial design of simple components one may safely put $\gamma_{f2} = 1.0$.

2. *Material factor* (γ_m). In the checking of *members*, BS.8118 adopts a constant value for this factor, namely $\gamma_m = 1.2$. For *connections*, the recommended value lies in the range 1.2–1.6, depending on the joint type and the standard of workmanship. Table 5.1 includes suggested values. The lower value 1.3 given for welded joints should only be used if it can be ensured that the standard of fabrication will satisfy BS.8118: Part 2. Failing this, a higher value must be taken, possibly up to 1.6.

It is emphasized that the Table 5.1 γ-values need not be binding. For example, if a particular imposed load is known to be very unpredictable, the designer would take γ_f higher than the normal value, or if there is concern that the quality of fabrication might not be held to the highest standard, γ_m ought to be increased.

Often a component is subjected to more than one type of action-effect at the same time, as when a critical cross-section of a beam has to carry simultaneous moment and shear force. Possible interaction between the different effects must then be allowed for. For some situations, the best procedure is to check the main action-effect (say, the moment in a beam) using a modified value for the resistance to allow for the presence of the other effect (the shear force). In other cases, it is more convenient to employ interaction equations. Obviously, a component must be checked for all the possible combinations of action-effect that may arise, corresponding to alternative patterns of service loading on the structure.

After checking a component for static strength, a designer will be interested in the actual degree of safety achieved. This can be measured in terms of a quantity *LFC* (load factor against collapse) defined as follows:

$$LFC = \frac{CR}{NA} \tag{5.3}$$

where *CR* and *NA* are as defined in Section 5.1.2. For a component which is just acceptable in terms of static strength ($FR = FA$), the *LFC* would be given by:

$$LFC = \gamma_f^* \gamma_m \tag{5.4}$$

where γ_f^* is the ratio of the action-effect under factored loading to that arising under nominal loading (i.e. a weighted average of γ_f for the various loads on the structure). Thus for example a typical member might have γ_f^* and γ_m equal to 1.3 and 1.2 respectively, giving a minimum *LFC* of 1.56. This implies that the member could just withstand a static overload of 56% before collapsing. The aim of limit state design is to produce designs having a consistent value of *LFC*.

Different results are obtained in checking static strength, depending on whether the Elastic or Limit State method is used. The two procedures may be summarized as follows (Figure 5.1):

1. *Elastic design*. The structure is analysed under working load, and stress levels are determined. These must not exceed an allowable stress, which is obtained by dividing the material strength (usually the yield or proof stress) by a *factor of safety* (FS). For slender members, the allowable stress is reduced to allow for buckling.
2. *Limit state design*. The structure is assumed to be acted on by factored (up) loading, equal to working loads each multiplied by a *loading factor*. It is analysed in this condition and a value obtained for the resulting 'action-effect' (i.e. axial force, moment, shear force, etc.) arising in its various components. In any component, the action-effect, thus found, must not exceed the factored (down) resistance for that component, equal to its calculated resistance divided by the *material factor*. By 'calculated resistance' is meant the estimated magnitude of the relevant action-effect necessary to cause failure of that component.

What really matters to the user of a structure is its actual safety against collapse. How much overload can it take above the working load before it fails? Safety may be expressed in terms of the quantity *LFC*. A sensible code is one providing a consistent value of *LFC*. Too high an *LFC* is oversafe, and means loss of economy. Too low an *LFC* is undersafe. By the very way it is formulated limit state design produces a consistent *LFC*. Elastic

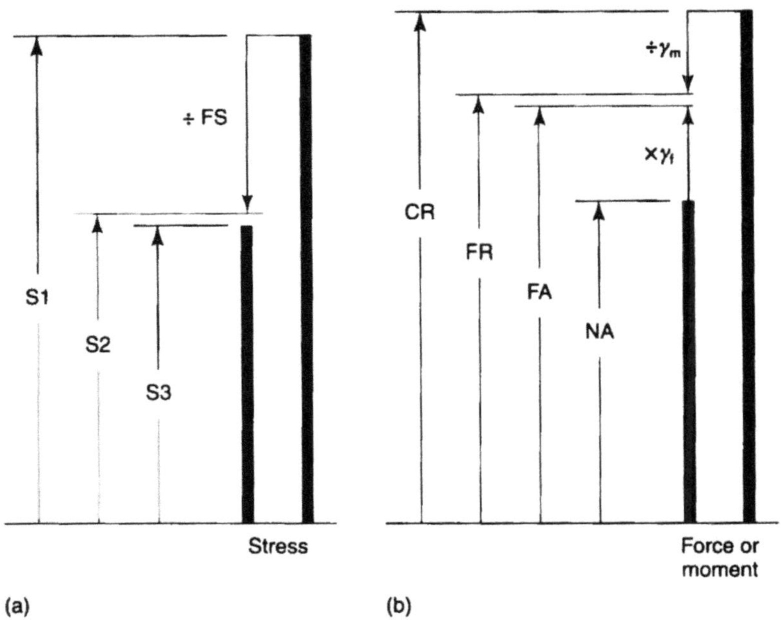

Figure 5.1 Static strength: (a) elastic design (S1 = material strength, S2 = allowable stress, S3 = stress arising at nominal working load); (b) limit state design.

design does not, because stress at working load is not necessarily an indication of how near a component is to actual failure.

5.1.4 Serviceability limit state

The reason for considering this limit state is to ensure that the structure has adequate stiffness, the requisite calculations being usually performed with the structure subjected to unfactored nominal loading. It is usually concerned with the performance of members rather than joints.

When a member is first taken up to its nominal working load, its deformation comprises two components: an irrecoverable plastic deflection and a recoverable elastic one (Figure 5.2). The main causes for the plastic deflection are the presence of softened zones next to welds (Chapter 6) and the rounded stress–strain curve. Further factors are local stress concentrations and locked-in stresses, which also lead to premature yielding (as in steel).

The serviceability check for a member simply consists of ensuring that its elastic deflection does not exceed an acceptable value:

$$\Delta_E \leqslant \Delta_L \tag{5.5}$$

where Δ_E = predicted elastic deflection under nominal loading, and Δ_L = limiting or permitted deflection.

A specific design calculation for the plastic deflection (under the initial loading) is never made. This is because it is usually small, and disappears on subsequent applications of the load. However, with materials having a very rounded stress–strain curve, the initial plastic deformation tends to be more pronounced, and there is a danger that it may be unacceptable. We cover this possibility in design by arbitrarily decreasing the limiting

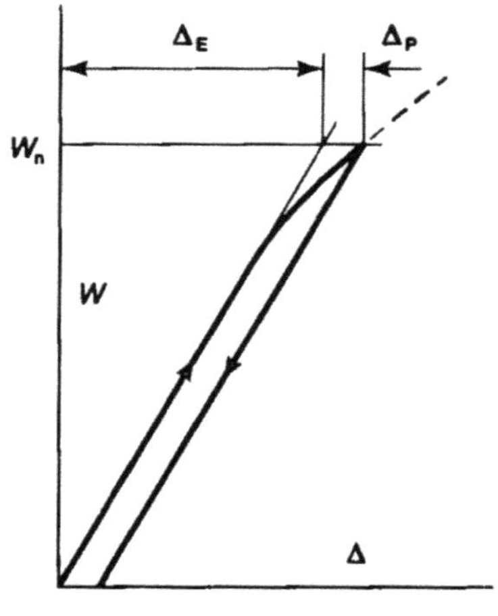

Figure 5.2 Elastic (Δ_E) and plastic (Δ_P) components of deflection at nominal working load.

stress for such materials, when checking the ultimate limit state (Section 5.3.1).

The type of member for which the serviceability limit state is most likely to be critical is a beam, especially if simply supported, for which Δ_E can be calculated employing conventional deflection formulae (Section 8.8). It is rarely necessary to check the stiffness of truss type structures.

The designer must decide on a suitable value for Δ_L, preferably in consultation with the client. The important thing is not to insist on an unduly small deflection, when a larger one can be reasonably tolerated. This is especially important in aluminium with its relatively low modulus. A general idea of the deflection that can be tolerated is given by the value suggested for purlins in BS.8118, namely $\Delta_L = \text{span}/100$ (under dead + snow + wind).

For a component that has to carry a combination of loads, the strict application of equation (5.5) may be thought too severe. A more lenient approach is to base Δ_E on a reduced loading, in which the less severe imposed loads are factored by γ_{f2} as given in Table 5.1.

Turning to joints, it is never necessary to check the deformation of welded ones, and even for mechanical joints an actual calculation is seldom required. With the latter, if stiffness is important, a simple solution is to specify close-fitting bolts or rivets, rather than clearance bolts. Alternatively, for maximum joint stiffness, a designer can call for friction-grip (HSFG) bolts, in which case a check must be made to ensure that gross slip does not occur before the nominal working load is reached. In so doing the calculated friction capacity is divided by a *serviceability factor* (γ_s), as explained in Section 11.2.7.

5.1.5 Limit state of fatigue

For a structure or component subjected to repeated loading, thousands or millions of times, it is possible for premature collapse to occur at a low load due to fatigue. This can be a dangerous form of failure without prior warning, unless the growth of cracks has been monitored during service.

Fatigue is covered in Chapter 12. The usual checking procedure is to identify potential fatigue sites and determine the number of loading cycles to cause failure at any of these, the design being acceptable if the predicted life at each site is not less than that required. The number of cycles to failure is normally obtained from an *endurance curve*, selected according to the local geometry and entered at a stress level (actually stress range) based on the nominal unfactored loading. Alternatively, for a mass-produced component, the fatigue life can be found by testing.

Table 5.2 Summary of limiting stresses needed for checking static strength

	Type of resistance calculation	Refer to
	Member design	
p_a	Localized resistance of section (axial force)	Table 5.3
p_o	Resistance of section to moment, or general yielding under axial load	Table 5.3
p_v	Resistance to shear force (beams)	Table 5.3
p_b	Resistance to overall buckling (beams, struts)	Section 5.4
	Joint design	
p_s	Resistance of fastener to shear	Section 11.1.4
p_p	Bearing resistance of plate	Table 5.3
p_t	Resistance of fastener to tension	Section 11.1.7
p_w	Resistance of weld metal	Section 11.3.4
p_f	Fusion boundary resistance at a weld	Section 11.3.5

Note. The final column shows where the relevant formula may be found.

5.2 THE USE OF LIMITING STRESSES

In order to obtain a component's calculated resistance, as required for checking the limit state of static strength, a designer needs to know the appropriate limiting stress to take. Table 5.2 lists the various kinds of limiting stress that appear in later chapters. The derivation of those needed in member design is explained in Sections 5.3 and 5.4. Those which arise in the checking of joints are covered in Chapter 11.

5.3 LIMITING STRESSES BASED ON MATERIAL PROPERTIES

5.3.1 Derivation

The first three limiting stresses in Table 5.2 must be derived from the quoted properties of the material and Table 5.3 gives suitable expressions

Table 5.3 Formulae for limiting stresses that depend on properties of member material

Type of resistance calculation	$f_u < 2f_o$	$f_u > 2f_o$
Local resistance of section (axial load)	$p_a = 0.5(f_o + f_u)$	$p_a = 1.5f_o$
Resistance of section to moment, or general yielding (axial load)	$p_o = f_o$	$p_o = 1.4f_o - 0.2f_u$
Resistance to shear force (beams)	$p_v = 0.6p_o$	
Bearing resistance of plate (mechanical joints)	$p_p = 1.1(f_o + f_u)$	
Fusion boundary resistance, welded joints	refer to Section 11.3.5	

Note. f_o, f_u = minimum values of 0.2% proof stress and tensile stress.

for so doing. In welded members, they are based on the strength of the parent metal, even though the material in the heat-affected zone (HAZ) is weakened due to the welding. The latter effect is looked after by taking a reduced ('effective') thickness in the softened region, as explained in Chapter 6.

The stress most used in member design is p_o. In steel codes, this is taken equal to the yield stress, and it seems reasonable in aluminium to employ the equivalent value, namely the 0.2% proof stress f_o. This is generally satisfactory, but problems can arise with 'low-n' materials having a very rounded stress–strain curve ($f_u \gg f_o$). For these, the use of $p_o = f_o$ will result in a small amount of irrecoverable plastic strain at working load, which may not be acceptable.

To allow for this, we propose that when $f_u > 2f_o$ a reduced value should be taken for p_o, as shown in Table 5.3. This expression has been designed to limit the plastic component of the strain at working load to a value of about 0.0002, i.e. one-tenth of the proof stress value (0.2%), assuming that the stress σ then arising is $0.65p_o$. Such an approach seems reasonable. The BS.8118 rule, which we believe to be over-cautious, is compared with ours in Figure 5.3. The estimated plastic strains at working load as given by the two methods are plotted in Figure 5.4, based on the Ramberg–Osgood stress–strain equation (4.3).

For the stress p_a, we generally follow the British Standard and take the mean of proof and ultimate stress (although some codes take $p_a = f_u$). Again, there is a problem with low-n material, although a greater degree of plastic strain can now be tolerated, as we are concerned with yielding at a localized cross-section of the member. In our method, we take a cut-off

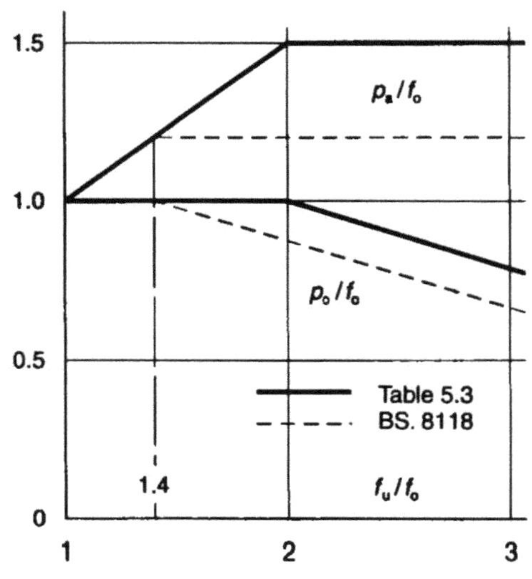

Figure 5.3 Relation between limiting stresses (p_o, p_a) and material properties (f_o, f_u).

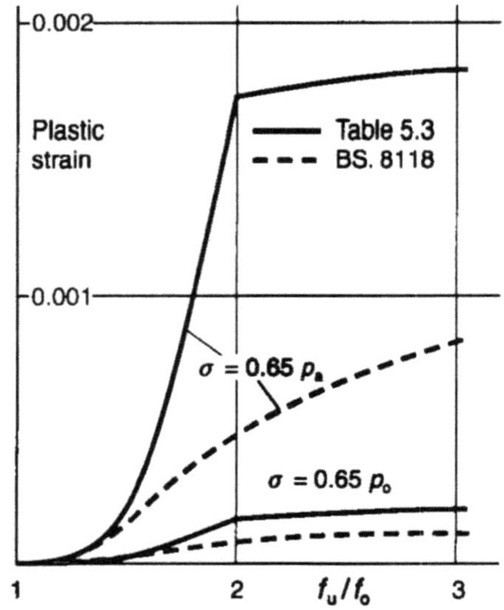

Figure 5.4 Plastic strain at working load, effect of ultimate/proof ratio.

at $2f_o$ (Table 5.3), which is aimed to limit the local plastic strain to a reasonable value of about 0.002, when $\sigma = 0.65p_a$.

It will be realized that the above procedures for dealing with low-n material are rather arbitrary. For some designs, the acceptable level of plastic strain may be lower than that assumed, while for others it may be higher. In such cases, the designer has the option of employing the stress–strain equation (4.3) to obtain more appropriate values.

The limiting stress p_v needed for checking shear force is based on the von Mises criterion in the usual way ($p_v = p_o/\sqrt{3} \approx 0.6p_o$).

5.3.2 Procedure in absence of specified properties

For some material, the specification only lists a 'typical' property and fails to provide a guaranteed minimum value, as for example with hot-finished material (extrusions, plate) in the 'as-manufactured' F condition. This creates a problem in applying the limiting stress formulae in Table 5.3. In such cases a reasonable approach is to take the property in question as the higher of two values found thus:

A. some percentage of the quoted typical value, perhaps 80% (the BS.8118 suggestion);
B. the guaranteed value in the associated O condition.

This is realistic for F condition material in extruded form, but can be unduly pessimistic when applied to plate. A typical example is plate in

5083-F, for which the actually measured 0.2% proof stress often greatly exceeds the specification value for 5083–0.

5.3.3 Listed values

Table 5.4 lists p_a, p_o and p_v for a selection of alloys, obtained as above. A designer may sometimes decide to deviate from these listed values. For

Table 5.4 Limiting design stresses for a shortlist of alloys

Alloy	temper	Form	Maximum t (mm)	Member design p_a (N/mm^2)	p_o (N/mm^2)	p_v (N/mm^2)	Joint design p_p (N/mm^2)	p_f (N/mm^2)
3103	0	S	50	52	31	19	137	62
	H14	S	25	130	120	72	286	62
	H18	S	3	175	165	99	385	62
5251	0	P, S	50	90	52	31	242	110
	H22	P, S	25	155	120	72	341	110
	H24	P, S	12.5	175	140	84	385	110
	H26	S	4	200	170	102	440	110
5154A	0	P, E	50	127	76	46	330	150
	H22	P	40	215	180	108	473	150
	H24	P	25	235	200	120	517	150
	H26	S	6	260	230	138	572	150
5083	0	P, S, E	50	187	120	72	440	200
	H22	P, S	40	260	215	129	572	200
	H24	P, S	25	295	250	150	649	200
6063	T4	E	25	97	65	39	214	97
	T5	E	3	152	130	78	335	106
	T5	E	3–25	135	110	66	297	94
	T6	E	10	192	170	102	423	125
	T6	E	10–25	177	160	96	390	115
6061	T4	P, S	12.5	157	110	66	346	157
	T4	E	25	145	110	66	319	145
	T6	P, S	12.5	265	240	144	583	172
	T6	E	25	250	240	144	550	162
6082	T4	P, S, E	12.5	157	110	66	346	157
	T5	E	5	250	230	138	550	175
	T6	P, S	6	285	260	156	627	185
	T6	P, S	6–12	277	255	153	610	180
	T6	E	5	270	250	150	594	175
	T6	E	5–25	285	260	156	627	185
7020	T4	P, S	12.5	265	210	126	583	265
	T6	P, S	12.5	315	280	168	693	283*
	T6	E	40	320	290	174	704	288*

Notes: 1. P = plate, S = sheet, E = extrusion.
2. *Increase by 5% when welding is under *strict* thermal control.

example, a reduced value might be thought desirable for extruded material stressed transversely (across the grain).

Table 5.4 includes values for the stresses p_p and p_f needed in joint design, which also depend on the member properties. See Chapter 11.

5.4 LIMITING STRESSES BASED ON BUCKLING

5.4.1 General form of buckling curves

The fourth limiting stress p_b relates to overall member buckling, for which we consider three possible modes:

LT beams in bending, lateral-torsional buckling;
C struts in compression, column (or 'flexural') buckling;
T struts in compression, torsional buckling.

Figure 5.5 shows the general form of the curve relating buckling stress p_b and overall slenderness parameter λ for a member of given material failing in a given mode. The 'curve' is shown as a scatter-band, since its precise position is affected by various sorts of imperfection that are beyond the designer's control. Curve E shows the ideal elastic behaviour that would theoretically be obtained assuming no imperfections, i.e. zero initial crookedness, no locked-in stress, purely elastic behaviour, etc. In the case of ordinary column buckling, this is of course the well-known Euler curve. The range of the real-life scatter-band in relation to curve E varies according to the mode of buckling considered. For design purposes, a curve such as D in Figure 5.5 is needed, situated near the lower edge of the appropriate scatter-band, with a cut-off at the limiting stress for the material.

Local buckling of a thin-walled cross-section, as distinct from overall buckling of the member as a whole, is covered in Chapter 7. This form of instability, which often interacts with overall member buckling, is allowed for in design by taking a suitably reduced ('effective') width for slender plate elements.

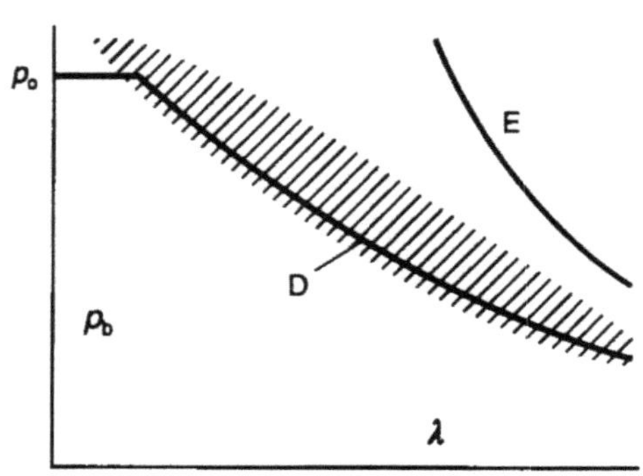

Figure 5.5 Variation of overall buckling stress with slenderness.

5.4.2 Construction of the design curves

It is convenient to represent buckling design curves by means of an empirical equation, containing factors that enable them to be adjusted up or down as required. Here we follow BS.8118 by employing the modified Perry formula, which is a development of the original Perry strut-formula that was devised by Ayrton and Perry in 1886. The modified Perry formula is a quadratic in p_b, the lower of the two roots being the one taken. It may be written as follows (valid for $\lambda > \lambda_1$):

$$(p_E - p_b)(p_1 - p_b) = cp_E p_b\left(\frac{\lambda}{\lambda_o} - \frac{\lambda_1}{\lambda_o}\right) \tag{5.6}$$

where: p_1 = intercept on stress-axis,
p_E = 'ideal' buckling stress (curve E) = $\pi^2 E/\lambda^2$
λ = slenderness parameter,
λ_o = intercept of plateau produced on curve E = $\pi\sqrt{E/p_1}$
λ_1 = extent of plateau on design curve,
c = imperfection factor,
E = modulus of elasticity = 70 kN/mm².

The solution to (5.6) is:

$$p_b = p_1\left(\frac{1-\sqrt{1-\alpha^2}}{\alpha x}\right) \tag{5.7}$$

where:

$$\alpha = \frac{2x}{x^2 + cx + 1 - c(\lambda_1/\lambda_o)}$$

$$x = \lambda/\lambda_o$$

Figure 5.6 shows the effect of c on the shape of the curve thus obtained, for given p_1 and λ_1.

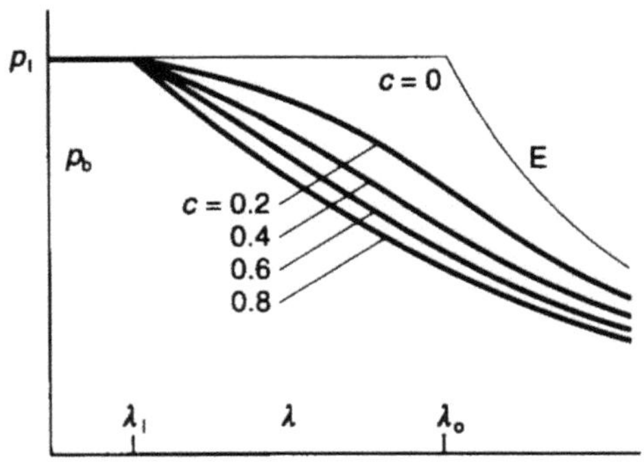

Figure 5.6 Modified Perry buckling formula, effect of c.

The parameter λ depends purely on the geometry of the member. For ordinary column buckling (C) it is simply taken as effective length over radius of gyration (l/r). For the other buckling modes (T, LT) it is defined as follows:

$$\lambda = \pi \sqrt{\frac{E}{p_E}} \tag{5.8}$$

where the ideal buckling stress p_E is as given by standard elastic buckling theory. In other words, λ is taken in such a way that the 'ideal' curve (E) is always the same as the Euler curve, whatever buckling mode is considered.

The stress p_1 at which the design curve intercepts the stress-axis would normally be put equal to the limiting stress p_o for the material (Section 5.3). However, a reduced value must be taken if the cross-section has reduced strength, due to either local buckling or HAZ softening at welds.

5.4.3 The design curves

The exact shape of the curve for a given value of p_1 is controlled by two parameters: the plateau ratio λ_1/λ_o and the imperfection factor c. In any buckling situation, these have to be adjusted to give the right shape of design curve. British Standard BS.8118 rationalizes overall buckling by adopting six series of design curves, with parameter values as given in Table 5.5. These are based on research by Nethercot, Hong and others [15–17]. The curves are compared in Figure 5.7 on a non-dimensional plot. Actual design curves of p_b against λ appear in Chapters 8 and 9, covering a range of values of p_1.

Table 5.5 Overall buckling curves – parameter values

Buckling case		Plateau ratio λ_1/λ_o	Imperfection factor c
Lateral-torsional buckling of beams	LT	0.6	0.10
Column buckling of struts	C1	0.2	0.20
	C2	0.2	0.45
	C3	0.2	0.80
Torsional buckling of struts	T1	0.4	0.35
	T2	0.6	0.20

Notes. 1. This table relates to equations (5.6), (5.7).
2. The resulting families of design curves are presented as follows: LT, Figure 8.22; C1, 2, 3, Figure 9.2; T1, 2 Figure 9.9.

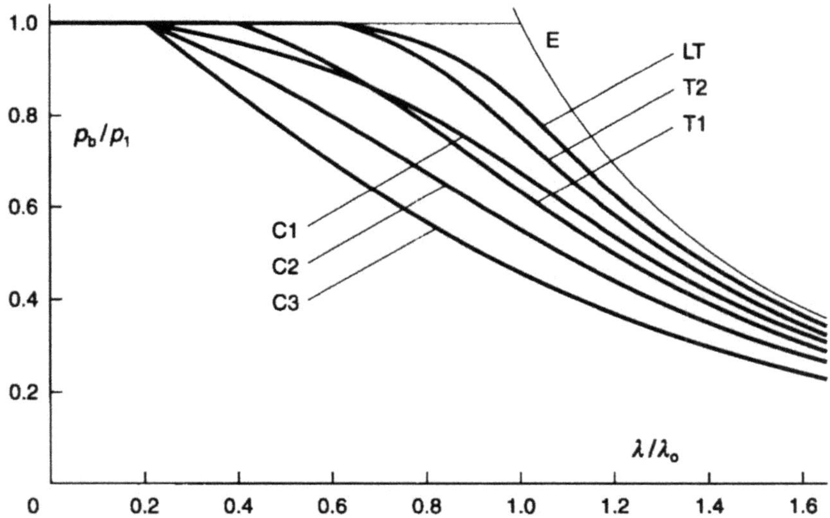

Figure 5.7 Comparison of the six buckling curves for overall buckling.

CHAPTER 6

Heat-affected zone softening at welds

6.1 GENERAL DESCRIPTION

An annoying feature in aluminium construction is the weakening of the metal around welds, known as HAZ (heat-affected zone) softening (Figure 6.1). With the 6xxx-series alloys, the heat of welding can locally reduce the parent metal strength by nearly half. With 7xxx alloys, the weakening is less severe, but extends further out from the weld. For work-hardened material (5xxx, 3xxx series), the metal in the HAZ becomes locally annealed, with properties falling to the O-condition level. Only for parent metal supplied in the annealed or T4 condition can HAZ effects be ignored.

The metal in the HAZ may be weaker than the actual weld metal, or it may be stronger, depending on the combination of parent and filler materials used. It often pays to locate welds in regions of low stress, i.e. away from the extreme fibres or at a section of low moment in a beam. A full-width transverse weld, as used for the attachment of a web stiffener, brings a high penalty as it causes softening right across the section.

A designer needs data on two aspects of HAZ softening: its severity and its extent. The severity is largely a function of the parent material used, while the extent depends on various factors. The ratio of the area of affected parent metal (transverse to the weld) to the weld deposit area can vary from 2 or 3 for a large multi-pass weld to 20 or more for single-pass.

It must be emphasized that the subject of HAZ softening is far from being an exact science. A well-known method for estimating the extent of

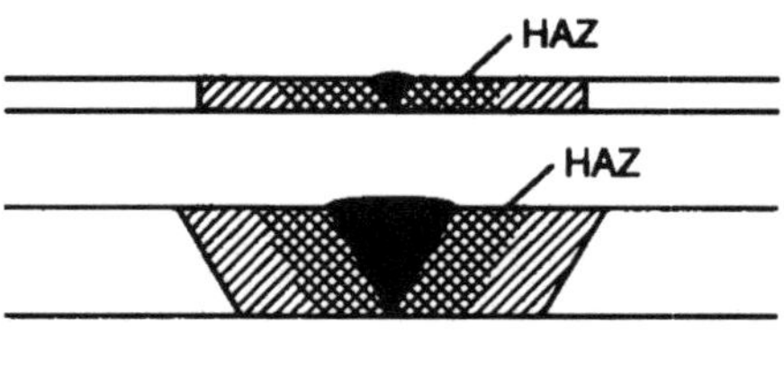

Figure 6.1 Zone of HAZ softening at aluminium welds.

the softening is the famous 'one-inch rule' which often proves adequate, but not always [18]. This simple method is explained in Section 6.5.3, after which we go on to present a more scientific treatment (RD method), which may be used to replace the one-inch rule in situations that demand a more accurate estimate of the HAZ extent (Sections 6.5.4–6.5.11).

Many people think that HAZ softening is such a minor effect that a very rough estimate of its extent is all that is needed. It is true that with large multi-pass welds, as used in massive members, the softened area only extends a short distance in relation to the size of the weld. For these, almost any extent-rule will usually do. But with smaller welds, as used in thin members, the extent of the HAZ is relatively much greater and a better approach is desirable. For these, the one-inch rule can lead to an estimate of member resistance that is unacceptably low.

The BS.8118 procedure for predicting HAZ extent gets the worst of both worlds, since it is more awkward to apply than the one-inch rule and often inaccurate. Our proposed method is more realistic, although still fairly approximate compared to most structural calculations. In recent years, special computer programs have become available, which accurately model the temperature changes during welding and the resulting metallurgical effect [19]. For a mass-produced component, it may be sensible to employ one of these. Alternatively, the HAZ pattern can be found experimentally by making a hardness survey on a prototype.

Designers should also be aware of the locked-in ('residual') stresses in welded components, even though these are not directly considered in the design process. As with steel, there is a region of locked-in longitudinal tensile stress at any weld, balanced by compressive stresses elsewhere in the section. But compared to steel, where the tensile stress at the weld is invariably up to yield, the stress levels in aluminium are relatively low. The zone of locked-in longitudinal tension at an aluminium weld is generally narrower than the HAZ. Mazzolani provides interesting plots of residual stress in welded aluminium members [26].

Most of this chapter specifically covers HAZ softening at welds made by the MIG process. TIG welds, for which the HAZ effects are much less predictable, are considered in Section 6.8. The friction-stir process is still very new, but some preliminary results suggest that the softening at FS welds will tend to be less extensive than that at arc welds (Section 6.9).

6.2 THERMAL CONTROL

The extent of the HAZ can be critically affected by the control of temperature during fabrication. In a large multi-pass joint, if no such control were exercised, the temperature of the surrounding metal would just keep on rising as more passes were laid, leading to a vastly enlarged area of softening. With 7xxx-type material, it would also increase the severity of the softening.

What matters is the temperature T_0 of the adjacent parent metal when any new weld metal is about to be deposited, known as the initial or interpass temperature. The following effects tend to increase T_0:

1. The metal is still hot from the welding of a nearby joint.
2. Insufficient cooling time has been allowed since the laying of previous passes in the same joint.
3. Preheat is used.
4. The ambient temperature is high, as in the tropics.

In order to limit the adverse effects of overheating, an aluminium fabricator is required to exercise thermal control, namely to ensure that T_0 never exceeds a specified maximum value. British Standard BS.8118 recognizes two levels of thermal control, normal and strict, as follows:

	Normal control	*Strict control*
6xxx, 5xxx-series alloys	$T_0 < 100°C$	$T_0 < 50°C$
7xxx-series alloys	$T_0 < 80°C$	$T_0 < 40°C$

All fabrication should satisfy normal control and this is what a designer would usually specify. With 6xxx or 5xxx-series material, there is often little advantage in going to strict control, since this affects the area of softening rather than the severity. There is a stronger case for strict control with 7xxx, as it also reduces the severity.

When in any doubt, design calculations should be based on normal control. The assumption of strict control can be justified only in the following cases:

1. a MIG-welded joint for which strict control is specified, with the maximum permitted value of T_0 stated to the fabricator;
2. an isolated joint containing one single-pass MIG weld laid without preheat.

It is obviously advantageous to be able to use the more favourable HAZ parameters corresponding to strict control, when possible, and it is necessary to be specific as to which joints can count as case (2). In our treatment we assume welds to be single-pass up to a size (w) of 8 mm (Section 6.5.5). The definition of an *isolated* weld is discussed in Section 6.5.10.

6.3 PATTERNS OF SOFTENING

6.3.1 Heat-treated material

Figure 6.2 shows patterns of softening at a single-pass MIG weld, as might be obtained with 6082-T6 and 7020-T6 material. Such plots are determined

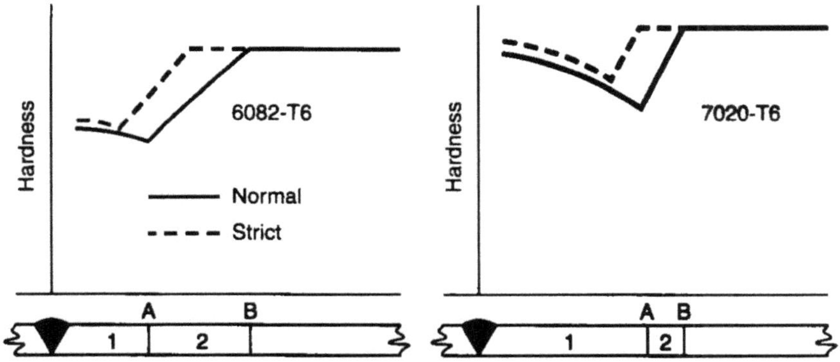

Figure 6.2 Typical hardness plots at a weld in heat-treated aluminium.

experimentally by conducting a hardness survey, and crudely indicate the variation in f_u (ultimate stress) rather than f_o (proof stress), since indentation hardness relates primarily to f_u. The plots in the figure have not been continued into the actual weld metal, because the strength of this varies with the filler used.

Two curves are shown for each alloy type, corresponding to welds made with normal and with strict thermal control. The HAZ can be divided into two regions (1, 2) as indicated on the plots for normal control. In region 1, the metal attains solution-treatment temperature and is thus able to re-age to some extent on cooling. In region 2, this temperature is not reached, and the metal is over-aged. The hardness is at a minimum at the boundary between the two regions (point A), and then rises steadily as we move out to point B. Beyond B, the heat of welding has negligible effect and full parent properties are assumed to apply. In region 1, the hardness increases as we move in towards the weld, although only slightly so for 6xxx material.

It is seen that the use of strict thermal control considerably reduces the extent of the softening for both alloy types. With 7xxx material, it also improves the properties in region 1 and reduces the amount of drop at A. With 6xxx material, the effect of thermal control on severity of softening is only slight. Very roughly, the relative widths of the two regions (1, 2) in Figure 6.2 satisfy the expressions below, where x_A and x_B are the distances of A and B from the middle of the weld (applicable to MIG welds):

$$\text{6xxx series} \qquad x_A \approx 0.5x_B \tag{6.1a}$$

$$\text{7xxx series} \qquad x_A \approx 0.8x_B \tag{6.1b}$$

6.3.2 Work-hardened material

With the non-heat-treatable alloys, HAZ softening need only be considered for work-hardened material; it is not a factor with extrusions or

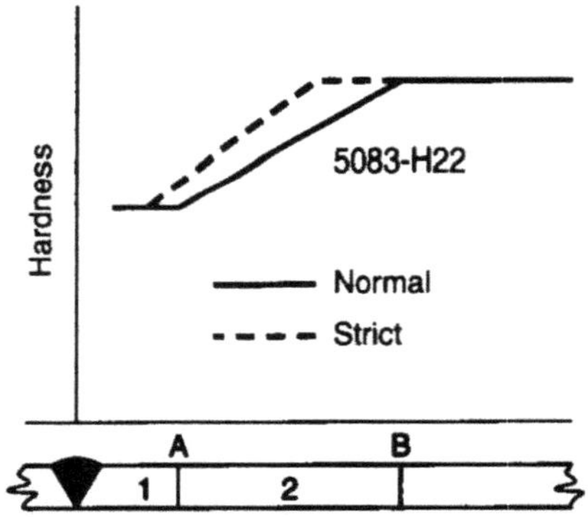

Figure 6.3 Typical hardness plots at a weld in work-hardened aluminium.

annealed plate. Figure 6.3 shows the pattern of softening that might be obtained at a single-pass MIG weld on 5083-H22 material. Again, regions 1 and 2 can be identified. In region 1, the hardness is now uniform and corresponds to the properties of the alloy in the annealed condition. As with the 6xxx series, the use of strict thermal control reduces the extent of the softening, but does not improve the strength in the HAZ. The relative widths of the two regions at a MIG weld are roughly given by:

$$\text{5xxx series} \qquad x_A \approx 0.3x_B \tag{6.1c}$$

A generally similar pattern would be obtained for 3xxx-series material, but possibly with a different width of softened area. Data on 3xxx materials are not generally available.

6.3.3 Stress–strain curve of HAZ material

Figure 6.4 compares typical stress–strain curves that might be obtained using coupons from the HAZ and from the parent metal. It is seen that the

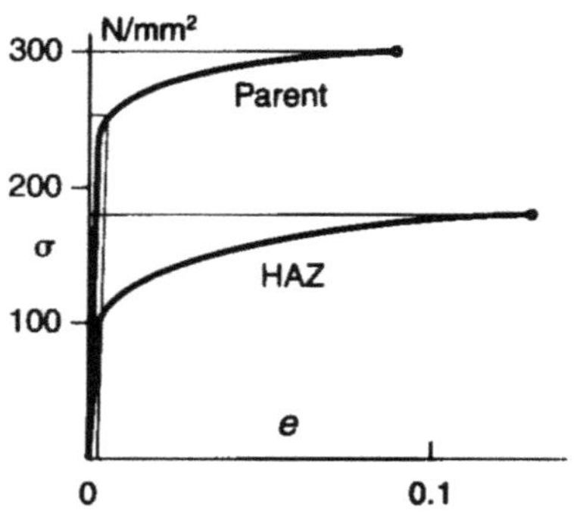

Figure 6.4 Parent and HAZ stress–strain curves compared (6082-T6).

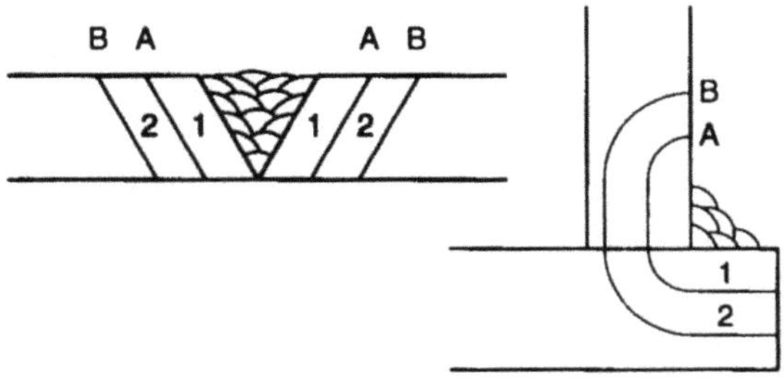

Figure 6.5 Pattern of softening at multi-pass weld on thick material.

HAZ curve has a more rounded knee, with a lower proof/ultimate ratio. Plots such as those in Figures 6.2 and 6.3, based on hardness surveys, give a visual picture of how the ultimate stress (f_u) is reduced in the HAZ. The drop in proof stress will be more marked. This is especially so for non-heat-treatable material (5xxx series) supplied in a hard temper.

6.3.4 Multi-pass welds

Figure 6.5 shows the typical softened zone at a large multi-pass weld. Regions 1 and 2 can again be identified, analogous to those shown in Figures 6.2 and 6.3 for a single-pass weld, now extending uniformly around the edge of the deposit. As we move away from the weld, the strength varies in the same general way as before, the lines A and B being metallurgically equivalent to points A and B in Figure 6.2 or 6.3.

6.3.5 Recovery time

With work-hardened alloys, the final HAZ properties are reached as soon as the metal has cooled after welding. But, with heat-treated material, the immediate strength in the HAZ is low, the final HAZ properties only being developed after enough time has elapsed to allow natural ageing to occur. Providing the component is held at a temperature of at least 10°C after fabrication, this time (the *recovery time*) may be roughly taken as:

6xxx-series alloys	3 days
7xxx-series alloys	30 days

If heat-treated material is held significantly below 10°C, the recovery time will be longer. On the other hand, quicker recovery can be achieved by post-weld artificial ageing. This involves holding the welded component at a temperature between 100 and 180°C for up to 24 hours, the exact procedure depending on the alloy. Such treatment also has a strengthening effect.

6.4 SEVERITY OF HAZ SOFTENING

6.4.1 Softening factor

The severity of softening in the HAZ is expressed in terms of a softening factor k_z which is intended to represent the ratio of HAZ strength to parent metal strength. At the current state of the art, it is only possible to suggest approximate values for this factor. Typically HAZ experiments employ hardness surveys, and although these give a good indication of the extent of the softened zone, they say much less about the actual tensile properties of the softened metal, because the hardness number correlates only crudely with tensile strength and hardly at all with proof stress. Also, there tends to be a lot of scatter between specimens.

In fact, for any given material we recognize three different values for k_z (as in BS.8118), and Section 6.6 explains which value to use when.

k_{z1} This value is used for calculations involving the limiting stress p_a (Table 5.2). Because $p_a = 0.5(f_o + f_u)$, the factor k_{z1} is notionally intended to represent the ratio of the HAZ value of this quantity to that for the parent metal, averaged over the width of region 1 (Figure 6.2).

k_{z2} This value is employed for resistance calculations that involve the limiting stress p_o. Because p_o is normally equal to the proof stress f_o, we (notionally) take k_{z2} as the ratio of HAZ proof to parent metal proof, again averaged over the width of region 1. The fact that the HAZ material has a lower proof/ultimate ratio than that for the parent metal (Figure 6.4) means that $k_{z2} < k_{z1}$.

k_{z3} This is a value which must be used for joint design in 7xxx material, when tensile stress acts transverse to the axis of the weld. It allows for the dip in HAZ properties at point A (Figure 6.2) and is notionally taken as the ratio of the quantity $0.5(f_o + f_u)$ at this point to its value for the parent metal. With 6xxx material and also non-heat-treatable material there is negligible dip at A, and $k_{z3} = k_{z1}$.

6.4.2 Heat-treated material

Table 6.1 lists proposed k_z-values for MIG-welded joints in 6xxx and 7xxx-series alloys. These have generally been pitched higher than the corresponding BS.8118 values, in line with current European thinking [20]. The 6xxx values are mainly based on results from specimens in the 6082 alloy, which are assumed to apply to any 6xxx material. Whether this is a valid assumption for the weaker kind of 6xxx alloy (such as 6063) is by no means clear.

Table 6.1 HAZ softening factor for heat-treated material at MIG-welded joints

Alloy series	Temper	k_{z1}	k_{z2}	k_{z3}
6xxx	T4	1.00	1.00	1.00
	T5	0.70	0.65	0.65
	T6	0.65	0.60	0.60
7xxx	T4	1.00	1.00	1.00
	T6	0.90	0.85	0.70
		(0.95)*	(0.90)*	(0.80)*

Note. *The second row of values of 7xxx-series material in the T6 temper (shown in brackets) may be used when strict thermal control is exercised during welding (see Section 6.2).

6.4.3 Work-hardened material

For work-hardened material, it may be assumed that the material in region 1 of the HAZ has the same properties as those for annealed material (Figure 6.3). Also, there is no dip at point A. The softening factor may therefore be taken as follows:

$$k_{z1} = k_{z3} = \frac{f_{oo} + f_{uo}}{f_o + f_u} \tag{6.2a}$$

$$k_{z2} = \frac{f_{oo}}{f_o} \tag{6.2b}$$

where f_o, f_u are the 0.2% proof stress and tensile strength of parent metal, and f_{oo}, f_{uo} are the same in the annealed condition.

6.5 EXTENT OF THE SOFTENED ZONE

6.5.1 General considerations

In the design of joints, one only needs to know the *severity* of softening in the HAZ. In member design, one must also know its *extent*, so that the total softened area at any critical cross-section can be determined. In discussing HAZ extent, two broad categories of welded joint may be recognized (Figure 6.6):

1. long straight joints, comprising one or more welds, as used for assembling a fabricated member from its component parts;
2. irregular attachment welds, as used for connections between members, or for securing local attachments, such as lugs, stiffeners, brackets, etc.

In massive members with multi-pass welds, category 1 joints cause softening in only a small proportion of the total section, and have a minor effect on the resistance. For these, a very approximate estimate of the HAZ extent is acceptable. But for small members, containing single-pass welds,

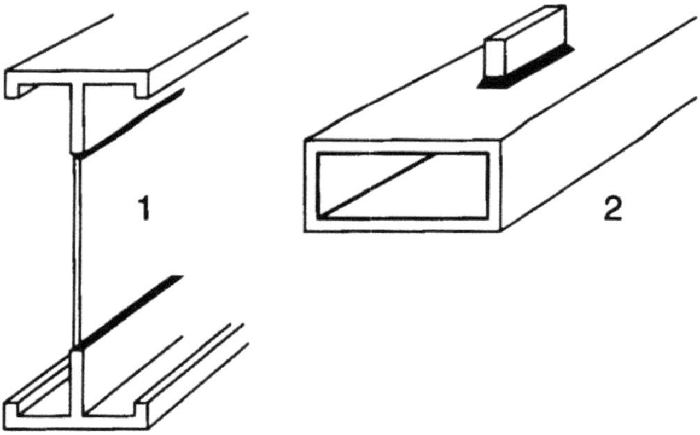

Figure 6.6 Categories of welded joint for estimation of HAZ extent: (1) long straight joints; (2) attachment weld.

the area of softening is relatively much greater and a more realistic estimate may be needed. Category 2 joints often extend over a large part of a member's width, causing a major proportion of the cross-section to become softened (or all of it).

The well-known method for estimating the extent of the HAZ is the *one-inch rule* [18]. Though crude, it is an invaluable design tool. We believe that a sensible strategy is to employ the one-inch rule for all preliminary calculations, with the option of switching to a more scientific method for the final check. In many cases it will be found that the effect of the HAZ on the resistance of the section is small, making the use of the one-inch rule quite acceptable. But in other cases, where preliminary calculations show that HAZ softening reduces the resistance significantly, worthwhile economies can be made by the use of a more refined treatment for the final design. In Sections 6.5.4–6.5.11, we present such a treatment, based on the work of Robertson at Cambridge during the 1980s [21], which leant heavily on the classic heat-flow equations of Rosenthal (Figure 6.7). We call this the 'RD method' [22].

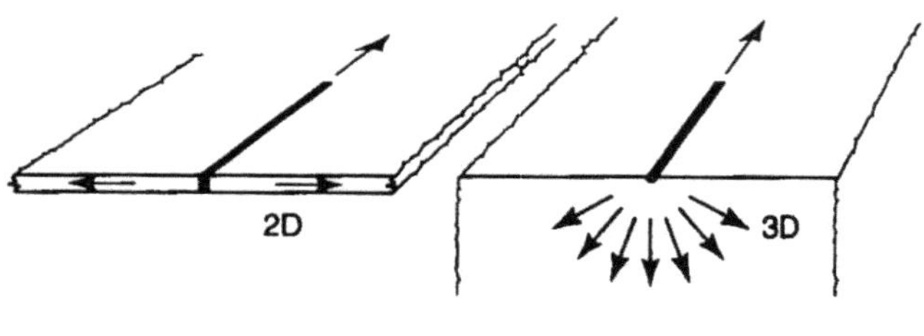

Figure 6.7 Heat-flow cases analysed by Rosenthal (moving heat source).

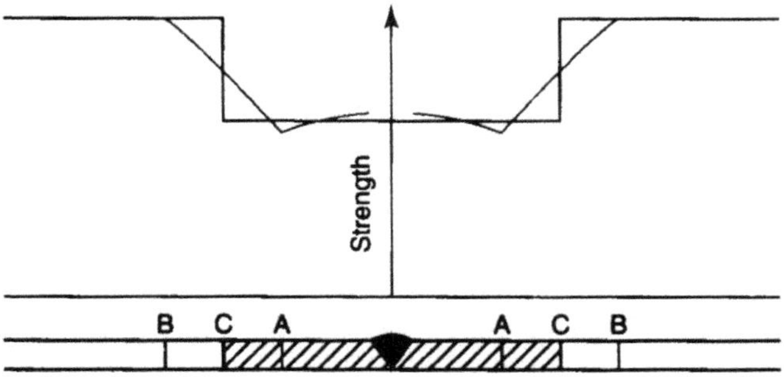

Figure 6.8 Extent (CC) of *nominal* HAZ for thin plate.

6.5.2 Nominal HAZ

In performing resistance calculations for welded members, the accepted practice is to use a *nominal* HAZ as an approximation to the true pattern of softening. In this, a weakened region of uniform strength is assumed adjacent to the weld, beyond which a step-change occurs to full parent strength.

Figure 6.8 illustrates the nominal pattern for joints in thin plate, with the step-change occurring at C, midway between the points A and B in the true pattern. A similar principle applies to joints in thick plate, with an assumed zone of uniform softening bounded by a line C (Figure 6.9).

6.5.3 One-inch rule

The one-inch rule was devised by Hill, Clark and Brungraber of Alcoa, and has been widely used since the 1960s. It simply states that the nominal HAZ extends a distance 1 inch (25 mm) in every direction from an appropriate reference point in the weld. For an in-line butt, the reference position is the centre-line of the weld (Figure 6.10(a)), while for a fillet it is at the root (6.10(b)).

With fillet welds on thick plate, the one-inch rule officially allows the HAZ boundary to be taken as an arc, as indicated in Figure 6.10(c). We believe this to be an unnecessary refinement that negates the simplicity of the rule, and instead would recommend the use of a square corner (as also

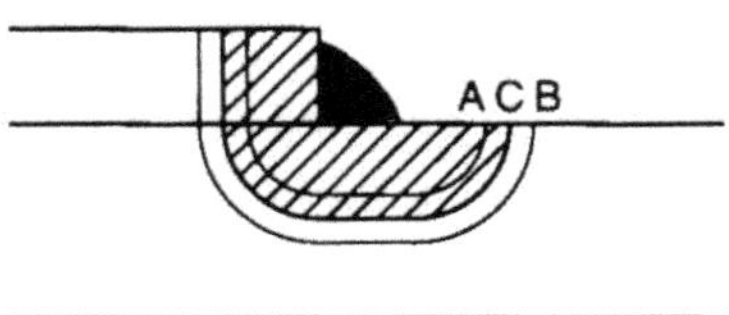

Figure 6.9 Extent of *nominal* HAZ (line C) for thick plate.

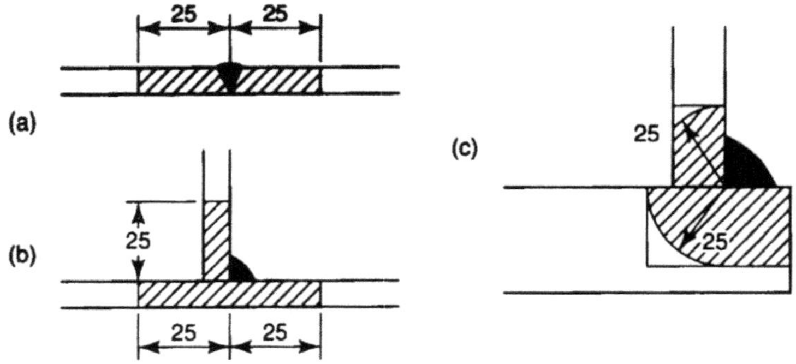

Figure 6.10 One-inch rule, extent of nominal HAZ.

shown). This makes for simpler calculations, the difference being small when compared to the overall degree of approximation in the method.

6.5.4 RD method

The one-inch rule is a brilliant simplification of a complex problem. Most of the time it works well and leads to acceptable predictions for the resistance of a welded member. But there are some situations where a lot of the cross-section gets softened, leading to a pronounced drop in resistance compared to the non-softened value. For these the one-inch rule can seriously exaggerate the extent of the HAZ, and hence significantly underestimate the performance of the member. In such cases, although the one-inch rule is convenient for use in preliminary calculations, it is clearly desirable to turn to a more scientific treatment for the final design [21, 22].

Below (Sections 6.5.5–6.5.11) we present such a treatment, referred to as our 'RD method'. This assumes that the nominal HAZ extends a distance s away from the weld deposit in every direction (Figure 6.11), where s is a function of weld size, alloy type and thermal control. In thin material, the boundary of the HAZ is taken straight across the thickness, as shown. In thick plate, when the HAZ only penetrates part way through, we take square corners instead of arcs for the sake of simplicity.

6.5.5 Weld geometry

For the purpose of our RD method, it is necessary to know the *effective* weld size w, and Figure 6.11 shows how we define this for some typical weld geometries. For any of these the relation between deposit area and weld size may be approximately taken as:

$$A_w = 0.7w^2 \qquad (6.3)$$

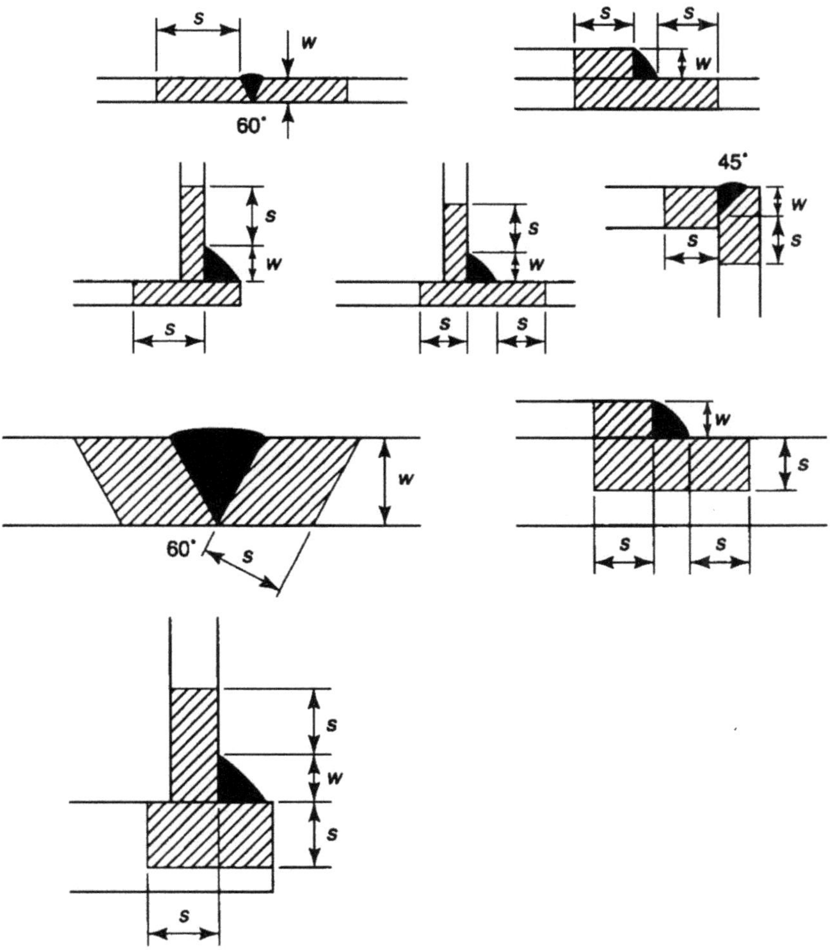

Figure 6.11 RD method, assumed HAZ geometries.

where A_w is the deposit cross-section, namely the actual added area including reinforcement or convexity. For other geometries, or preparation angles, the effective weld size w should be taken thus:

$$w = 1.2\sqrt{A_w} \tag{6.4}$$

where A_w includes a realistic allowance for convexity of the weld profile. If, at the design stage, the weld has not yet been detailed, w must be estimated from a knowledge of the plate thicknesses, a liberal value being assumed so as not to underestimate the HAZ.

With small welds, it is easy for the welder to lay a larger deposit than that shown on the drawing, leading to an increased area of HAZ. This can be allowed for in design by replacing the specified weld size w by a modified value w_1 obtained thus:

$$w_1 = \text{lesser of } \alpha w \text{ and } 8 \text{ mm} \tag{6.5}$$

where α is an oversize factor. The factor α should typically lie in the range 1.0–1.1 for butt welds and 1.0–1.2 for fillets, depending on the standard of deposit control exercised in fabrication. The HAZ at a small weld can grow by 40 or 50% with an over-enthusiastic welder.

We assume that MIG welds will be single-pass up to $w = 8$ mm, corresponding to $A_w = 45\ \text{mm}^2$. Beyond this they are treated as multi-pass welds. Also, our treatment assumes that for large welds the deposit area per pass is $45\ \text{mm}^2$. In practice, the maximum deposit area per pass is usually less than this, perhaps $40\ \text{mm}^2$, and our assumptions are therefore slightly pessimistic in that they tend to overestimate the extent of the HAZ.

6.5.6 Single straight MIG weld

First we consider the basic case of a long straight MIG-welded joint containing a single weld (Section 6.5.1, category 1). Our method is an attempt to rationalize Robertson and Dwight's results [21], three ranges of weld size being recognized: small, medium, big. The rules for small and big welds are closely based on Robertson's research, while for the medium range we employ an empirical procedure to bridge the gap.

It is found that a radically different approach is needed for small welds, as compared with big ones [22]. With small welds, the operative quantity is the area of the HAZ, this being a simple multiple of the area of the deposit. For such welds, therefore, the dimension s increases with size of weld; also, it varies according to the number of heat-paths away from the weld. In contrast, at big welds, s is unaffected by weld size or joint geometry.

1. *Small welds* ($w \leqslant 8$ mm). In this range it is assumed that the weld is single pass, and that Rosenthal's two-dimensional case is relevant (Figure 6.7). The area of the HAZ is found directly as follows:

$$A_z = KA_w = 0.7\,Kw_1^2 \tag{6.6}$$

where: A_z = section area of nominal HAZ,
A_w = actual section area of weld deposit (added area),
w_1 = modified weld size (equation 6.5),
K = factor to be read from Table 6.2.

The actual distribution of the HAZ material can then (if necessary) be found by determining s so as to bring A_z to the value obtained above.

2. *Medium welds* ($8 < w < 20$ mm). Two estimates are made using the

Table 6.2 HAZ-extent parameters for a single straight MIG weld

Alloy group:		5xxx, 6xxx		7xxx	
Thermal control: $T_0 <$		Strict 50°C	Normal 100°C	Strict 40°C	Normal 80°C
Weld size:					
Small ($w < 8$ mm)	$K =$	10	14	14	18
Medium ($8 < w < 20$):					
Rule P	$A_z =$	448 mm^2	627 mm^2	627 mm^2	806 mm^2
Rule Q	$s =$	$0.7w$	$0.85w$	$0.85w$	w
Big ($w > 20$ mm)	$s =$	14 mm	17 mm	17 mm	20 mm

empirical rules P and Q, and the less favourable then taken:

Rule P: A_z is put equal to the relevant value in Table 6.2, namely that which equation (6.6) would give when $w = w_1 = 8$ mm.

Rule Q: Dimension s is taken equal to the appropriate value in Table 6.2 (proportional to the weld size).

3. *Big welds* ($w > 20$ mm). The reasoning in this range is that each pass has a 'zone of influence' as typically shown in Figure 6.12, which can be related to Rosenthal's three-dimensional heat-flow case (Figure 6.7). The radius s of this zone depends on the alloy type and the level of thermal control. Table 6.2 gives proposed values of s pessimistically based on a deposit size of 45 mm^2 per pass. The nominal extent of the overall HAZ is constructed by taking the envelope of the zones of influence for all the passes. This is achieved by drawing a series of lines, as shown, each distant s from the relevant welded surface, arcs being replaced by square corners.

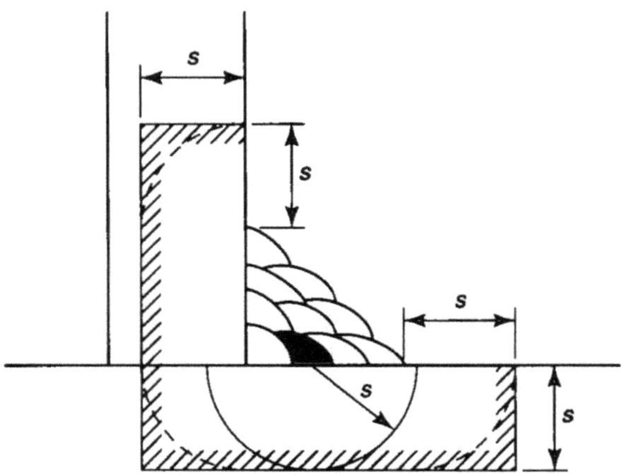

Figure 6.12 Assumed extent of nominal HAZ at a large multi-pass weld, showing 'zone of influence' for one of the passes.

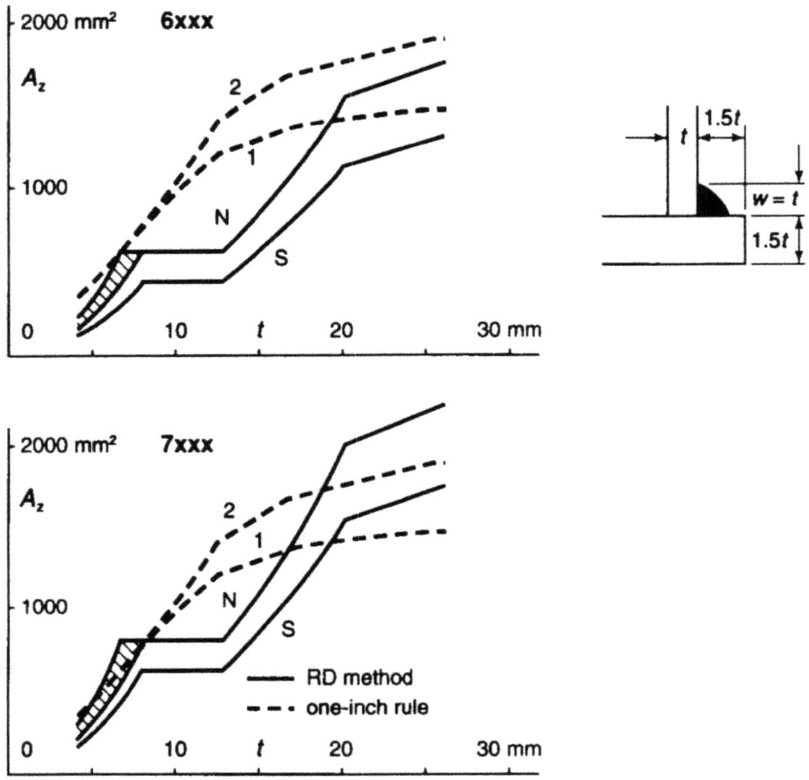

Figure 6.13 Predicted area (A_z) of nominal HAZ, comparison of RD method and one-inch rule. RD method: N = normal thermal control; S = strict thermal control. One-inch rule: curve 1 assumes arcs; curve 2 assumes square corners.

6.5.7 Variation of HAZ-extent with weld size

The upper plot in Figure 6.13 shows how the area A_z of the nominal HAZ, as predicted in Section 6.5.6, varies with size for a long straight fillet weld (MIG) of typical geometry at a joint in 6xxx-series plate. Two curves (N, S) are plotted, corresponding to normal and strict thermal control. On curve N, a hatched area appears in the region $t < 8$ mm to indicate how A_z is affected by deposit control ($\alpha = 1.0$–1.2, see Section 6.5.5). For curve S, which relates to strict thermal control, it is assumed that the deposit control would also be strict ($\alpha = 1$). Note that, when $t < 8$ mm, curve S is valid even if strict thermal control has not been specified, provided the joint is far enough away from other joints to count as 'isolated'.

Comment is needed on the fact that there is a region on each curve in which A_z is shown constant and not going up with weld size. Referring to either curve, the point defined by $t\,(=w) = 8$ mm represents the largest

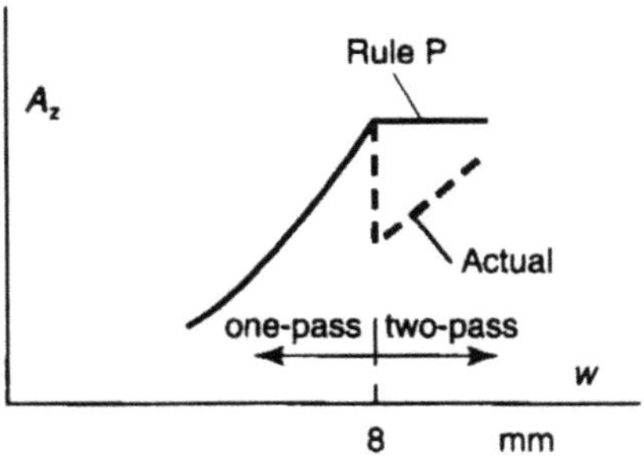

Figure 6.14 Variation in A_z for weld sizes around w = 8 mm for the joint geometry in Figure 6.13.

size of single-pass weld. A slightly bigger weld (say $w = 9$ or 10 mm) will be laid in two passes and have a smaller softened area than that for a single-pass weld with $w = 8$ mm. This is because the deposit area per pass will be less than for the 8 mm weld, coupled with the fact that there will be a high degree of overlap between the zones of influence for the two passes. The true plot of A_z versus weld size therefore dips suddenly at $w = 8$ mm and then climbs, as illustrated in Figure 6.14 (broken line). This would be hard to codify, and for design we conservatively draw a horizontal line as shown (rule P).

The estimated area of nominal HAZ for welds on 5xxx-series material was shown by Shercliff [23] to be the same as that for 6xxx, even though the actual pattern of softening is different. The 6xxx curves therefore apply to 5xxx series too.

Figure 6.13 also includes equivalent curves for welds on 7xxx-series material. It is seen that A_z is generally higher than for 6xxx material, the 7xxx-series S-curve coinciding with the 6xxx-series N-curve.

Included in Figure 6.13 are curves obtained by applying the one-inch rule to this geometry. It is seen that, for smaller welds (say $w = 4$–15 mm), the one-inch rule can seriously overestimate the HAZ-extent for 5xxx and 6xxx material, but gives a more reasonable answer for 7xxx material. At the worst, the one-inch predictions for welds on 6xxx material are too high by a factor of 2.5 or 3. This is the range where reasonable accuracy is most desirable. For big welds ($w > 20$ mm), the one-inch rule overestimates A_z for 6xxx material and gets it about right for 7xxx material. However, in this range a very approximate answer is acceptable.

6.5.8 Overlapping HAZs

It often happens that the individual HAZs due to two different welds overlap, thus producing a single softened zone (Figure 6.15). This occurs in a multi-weld joint, and also when two joints are near together. The combined HAZ for the two welds is found by plotting the extent of the

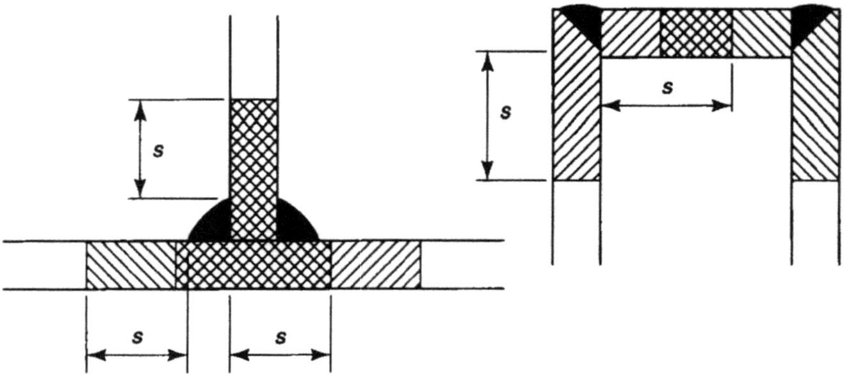

Figure 6.15 Overlapping HAZs.

nominal HAZ for each in turn, using the appropriate dimension *s* for each weld treated on its own, and then taking the enclosed area. The overlap area is counted just once.

This procedure will no longer be valid when simultaneous deposits are laid in the two welds, using a twin-torch technique. This may lead to a greatly increased HAZ.

6.5.9 Attachment welds

The basic method in Section 6.5.6 was developed from a study of long straight welds. The case of localized 'attachment welds' (Section 6.5.1, category 2) is more difficult.

Referring to Figure 6.16, the aim is to find the area of the HAZ material at a critical cross-section of the main member. This extends the full width of the connected part plus a projection *s* at each side, or up to a free edge if this is nearer. The HAZ penetrates a distance *s* into the main member, or the full thickness *t* if this is less. If the weld passes close to a corner of the section, the HAZ extends around the corner to give the same total area of HAZ. When the weld is circumferential, as in the attachment of a tubular branch member, the HAZ should be taken as extending right across the enclosed region.

The problem is to decide on a suitable value of the parameter *s* for such

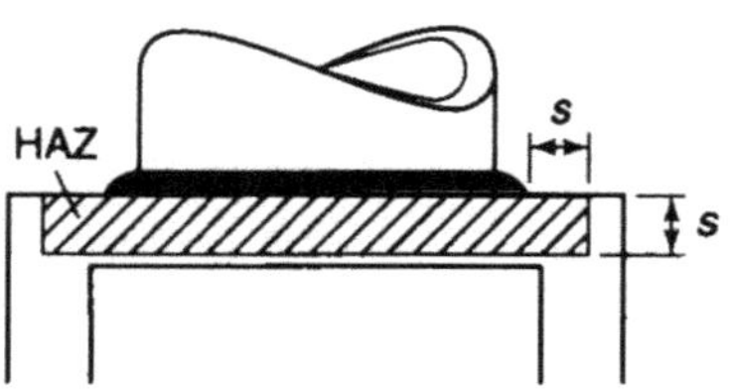

Figure 6.16 Nominal HAZ at an attachment weld.

welds. An instinctive procedure might be to consider a section transverse to the axis of the weld, and determine s in the same way as for a long straight joint of the same geometry, using the data in Table 6.2. This is not acceptable. In practice, the whole of an attachment weld, or all of one pass, will be laid non-stop without pause. The resulting heat flow, which is complex, would have to be determined on a one-off basis using a specialist computer program or else experimentally, neither of which may be practicable. The only sure thing is that the heat flow, and hence the pattern of softening, will be more unfavourable than that based on a straight weld.

The designer, therefore, should first calculate the value of s for an equivalent long straight weld of the same size, and then increase this by a suitable factor to allow for all the imponderables. For lack of hard data, we would suggest an arbitrary increase of 25% in the value of s, although this may be pessimistic when applied to a very short attachment weld (connection of a small lug, say).

With a wide attachment, the precise amount that the HAZ projects at either side is relatively unimportant, since it forms only a small part of the total affected width. The penetration into the thickness is more critical. If a small attachment weld is used on a relatively thick member, the HAZ penetrates only part way into the thickness, leading to a much smaller area of HAZ than that based on full penetration such as would be predicted by the one-inch rule.

6.5.10 Definition of an isolated weld (10A-rule)

For an isolated single-pass weld, it is permissible to use the more favourable HAZ parameters based on strict thermal control, even when only normal control is specified. How do we define 'isolated'? The following '10A-rule' for so doing is derived from Rosenthal's heat-flow equations [22]. Referring to Figure 6.17, weld Y may count as isolated provided that:

$$A_p \geqslant 10A_{zp} \tag{6.7}$$

where A_p = cross-section area of plate on the heat-path from any neighbouring weld (X) to the weld considered (Y), as measured from the nearest point in each weld, and A_{zp} = section area of that part of the nominal HAZ at weld X that lies in the interconnecting plate, measured from the nearest point in weld X.

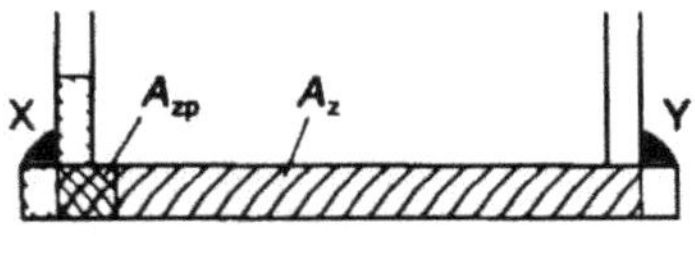

Figure 6.17 Check for an isolated weld, '10A-rule'.

In determining A_{zp}, an appropriate level of thermal control should be assumed for weld X. When a member contains just two welds, both may count as if welded under strict control, provided equation (6.7) is satisfied and both are single-pass.

6.5.11 RD method, summary

The procedures described in Sections 6.5.4–6.5.10 are summarized below. They enable the area and extent of the nominal HAZ (Section 6.5.2) to be found at any given cross-section in a member, adjacent to a MIG-welded joint.

1. Decide on the weld size w, or the modified value w_1 in the case of a small weld (Section 6.5.5), or estimate w from the plate thickness.
2. Decide on the effective level of thermal control (normal or strict) (Section 6.2 and 6.5.10).
3. Decide whether the joint is category 1 (long straight) or category 2 (attachment weld), see Section 6.5.1.
4. *Category 1* joint containing a single MIG weld. Figure 6.11 shows assumed shape of nominal HAZ, defined by parameter s. Section 6.5.6 explains how to find s, using Table 6.2.
5. *Category 1* joint containing two or more MIG welds. Obtain s as in 4 for each weld treated on its own, and then proceed as in Figure 6.15. (Section 6.5.8).
6. *Category 2* joint. Base the nominal HAZ on a value of s which is 25% higher than that for a category 1 joint having the same weld section geometry. See Section 6.5.9.

6.6 APPLICATION OF HAZ DATA TO DESIGN

6.6.1 Design of members

Some members are welded in such a way as to cause total softening at a particular cross-section, as for example at the end connection of a tubular member in a welded truss, or where transverse stiffeners are welded to an I-beam. In such cases, the resistance of the cross-section is simply found by making a conventional calculation, but using the appropriate limiting stress ($k_z p$) in place of the parent value (p).

Such a calculation would be pessimistic if applied to cross-sections that are only partially softened, as it would leave the non-softened material understressed. Instead, the accepted practice for these is to take an *effective section*, in which HAZ softening is allowed for by notionally removing some of the area locally around each weld. The resistance is then based on the effective section, instead of the actual one, assuming it to be entirely composed of full strength (unsoftened) material.

The effective section is obtained by replacing the area A_z of the nominal HAZ in each softened region by an effective area $k_z A_z$. The effective section properties may then be found by one or other of the following methods (refer also to Sections 10.2.4, 10.3.4):

1. *Method 1.* An effective plate thickness of $k_z t$ is generally assumed in each nominal HAZ region, instead of the true thickness t. In thick material, such that the nominal HAZ penetration s is less than t, the effective thickness is taken equal to $\{t - s(1 - k_z)\}$. The properties of the effective section thus obtained are then determined in the usual way.
2. *Method 2.* The section properties are first calculated with HAZ softening ignored, and suitable deductions then made corresponding to a 'lost area' of $A_z(1 - k_z)$ at each HAZ.

Method 2 is generally preferred for members just containing small longitudinal welds, since it only involves A_z which is readily found (equation (6.6)), without the need to know the exact location of the softened metal.

For any given material there are in general three possible k_z-values (see Section 6.4.1). The first two of these are needed for member design and are selected as follows:

k_{z1} shear force resistance in beams (Section 8.3);
local failure in tension and compression members (Section 9.3).

k_{z2} moment resistance in beams (Section 8.2);
general yielding in tension members (Section 9.4);
overall buckling (Sections 8.7.4, 9.5, 9.6).

In service, the strains arising in the softened zones at a partially welded cross-section are the same as those in the adjacent unsoftened material. With the above method of calculation therefore, using an effective section, the HAZ material will yield prematurely – well before the calculated resistance of the member is reached, and possibly even at working load. Some inelastic deflection therefore takes place, leading to a slight permanent set after the first loading. This does not matter, since the component will behave elastically under further load applications, and the inelastic deflection is small. (A similar state of affairs exists with steel members, where premature yielding occurs in parts of a section due to the very considerable locked-in stresses present in all steel components – much worse than in aluminium.)

6.6.2 Design of joints

In determining the resistance of a load transmitting joint, two calculations are generally needed: one for failure of the actual weld metal, and one for failure in the fusion zone – see Section 11.3. The latter refers to a failure plane in the HAZ, close to the weld deposit, for which the limiting stress is taken as the basic parent value factored by k_{z1}.

With welds in 7xxx material, carrying transverse tension, a check is also needed for possible failure at the dip in HAZ (point A in figure 6.2). For this we take $k_z = k_{z3}$.

6.7 COMPARISON WITH ONE-INCH RULE

It has already been stated that the most useful HAZ treatment for general use in design is the simple, but conservative, one-inch rule. In many situations, the softened region is relatively small and any slight over-estimate of its extent, through employing the one-inch rule, is justified by the simplicity. But there are other cases where HAZ effects cause a more serious drop in the resistance of a member, perhaps 10 or 20%. For these, the one-inch rule becomes too pessimistic, and something better is then needed if an economic section is to be arrived at. Our RD method is proposed as a more scientific alternative.

Figure 6.18 shows two beam sections for which relevant parameters are as follows:

Alloy	6082-T6
Welding process	MIG
Limiting stress p_o	260 N/mm^2
Softening factor k_z	0.60.

Included against each section is a bar-chart showing how the ratio M_c/M_{cg}

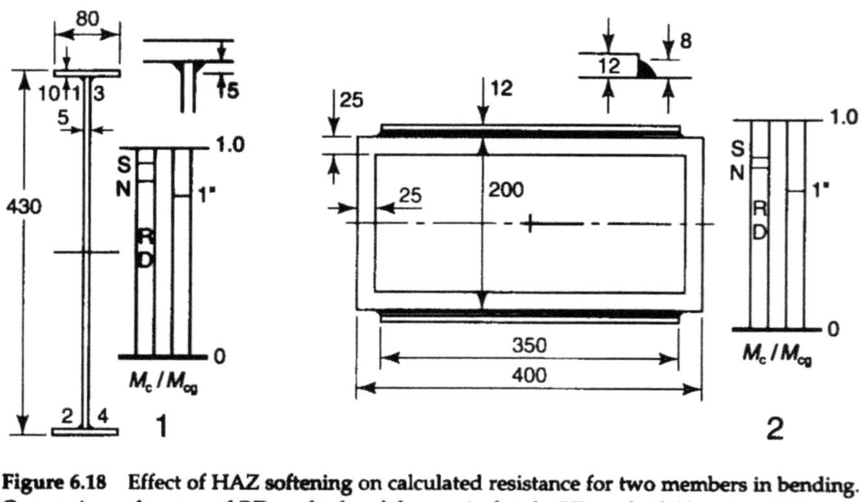

Figure 6.18 Effect of HAZ softening on calculated resistance for two members in bending. Comparison of proposed RD method and the one-inch rule. RD method: N = normal thermal control; S = strict thermal control.

varies, depending on which HAZ method is used, where M_c = calculated moment resistance with softening allowed for, and M_{cg} = moment resistance based on gross section properties (HAZ effects ignored).

Section 1 is a conventional plate-girder, with a 5 mm web that is just semi-compact. In applying the RD method, we cover two possibilities S and N as follows:

S Strict thermal control, with strict deposit control ($\alpha = 1.0$);
N Normal thermal control, and a more liberal deposit control ($\alpha = 1.2$).

The predicted fall in resistance below the unsoftened value is as follows:

RD method:	S	8%;
	N	16%;
One-inch rule:		23%.

It is worth noting that strict control (S) is quite realistic for a section such as this. With the welding sequence shown, it is only necessary for strict control to allow cooling between welds 2 and 3 (i.e. when the job is turned over to weld the second side). The 10A-rule shows that there is no need for a cooling pause after weld 1 or 3.

Section 2 is a large extruded box-section with square pads welded on top and bottom, the member being bent about its minor axis. The HAZ effect is considerable, because a large proportion of the top and bottom flanges is softened. In applying the RD method, we again cover strict and normal thermal control (S, N), and in so doing include the 25% increase in s due to the welds being category 2 (attachment welds). Deposit control is not an issue, because the weld size is the maximum for single-pass. The predicted losses in resistance due to HAZ effects are as follows:

RD method:	S	19%;
	N	23%;
One-inch rule		34%.

Even when strict thermal control is not specified, the S-value will still be generally valid, because the welds are far enough apart to count as 'isolated' when the 10A-rule is applied. This would only cease to apply if preheat were used, or there was other welding nearby.

6.8 HAZ AT TIG WELDS

6.8.1 Difference between TIG and MIG welding

The data provided in this chapter have so far assumed the use of the MIG-welding process. When TIG welding is used, a common choice for small welds, the situation changes.

With MIG welding, an electrode-positive DC process, a steady two-thirds of the heat input to the arc is used in melting off electrode wire, while the other third is released at the weld pool. The total heat input therefore, and hence the extent of the HAZ, is proportional to the size of the weld deposit. This forms the basis of equation (6.6). With TIG welding, an AC process, the total arc energy is about equally shared between the tungsten electrode and the weld pool. The heat released at the weld pool is thus half the total energy input. However, only part of this goes into the (hand-held) filler wire and the amount available for melting off this latter is less than half the total, perhaps only a third or a quarter, as compared with two-thirds for the MIG process. For a given size of deposit (equal to the amount of filler consumed), the extent of the HAZ (a function of the total energy) is therefore considerably greater for TIG than it is for MIG welding.

A further factor is that the TIG process is more dependent on the operator's technique, as compared with the semi-automatic MIG process. If the torch is allowed to dwell, the heat input per unit length of weld goes up, causing the extent of the HAZ to increase still further locally. The result is that the HAZ extent at a TIG weld is (a) greater than for a MIG weld having the same deposit size, and (b) variable.

Unfortunately, no systematic study appears to have been made of HAZ effects at TIG welds, and we only have a limited number of hardness surveys to refer to. These confirm that the extent of the softening is much more than with MIG. More information is needed.

6.8.2 Severity of softening with TIG welding

With non-heat-treatable material (3xxx, 5xxx series), it is reasonable to assume that the use of TIG welding will not increase the severity of HAZ softening, only its extent. Therefore, the k_z-values given in Section 6.4.3 remain valid.

With 6xxx-series material, we have seen that the severity of softening for MIG welds is only slightly affected by the level of thermal control, and one may reasonably infer that the values in Table 6.1 could also be used with TIG welding. Data from a very limited range of test results, however, have been interpreted as showing that such a view is optimistic and that lower k_z-values should be taken.

With 7xxx-series material, it is known that the severity of softening, especially the dip at the point of minimum strength (Figure 6.2), is significantly more pronounced with normal thermal control than it is with strict. This suggests that with the increased heat input of TIG the softening will be definitely more severe.

Summarizing, we would at present recommend that k_z-values be taken as follows when welds are made by TIG:

3xxx, 5xxx	same as for MIG;
6xxx	reduce MIG value by 0.05;
7xxx	reduce MIG value by 0.2.

6.8.3 Extent of the softened zone for TIG welding

With the present dearth of information, we would tentatively suggest that the extent of the nominal HAZ at a TIG weld should be found by making the following arbitrary adjustment to the predicted extent for MIG:

- *Welds on thin material.* Increase A_z by 50%.
- *Welds on thick material.* Increase s by 25%.

A problem arises when an original MIG weld is repaired using TIG. This will locally increase the width of the HAZ by an amount that is very hard to quantify.

6.9 HAZ AT FRICTION-STIR WELDS

At the time of writing, it is too soon to provide comprehensive HAZ data for joints made by the new friction-stir process. However, from a study of some hardness surveys provided by Hydro-Aluminium in Norway, using 6xxx-series specimens, it is possible to make the following observations for material welded in the T6 condition.

1. The pattern of softening will be of the same general form as that shown in Figure 6.2.
2. The severity of softening is comparable to that with MIG.
3. The extent of the softened zone relates to the shoulder radius of the welding tool.
4. The softened area will tend to be rather less than that obtained with MIG.

Note, however, that it is possible to FS-weld together extruded sections in the T4-temper, and then apply post-weld ageing to bring them up to full T6 properties. Then there is no HAZ region worth speaking of and the ductility is good. Such a procedure is readily carried out in an extrusion plant where long ageing furnaces already exist.

CHAPTER 7

Plate elements in compression

7.1 GENERAL DESCRIPTION

7.1.1 Local buckling

In aluminium design it is often economic to employ wide thin sections, so as to obtain optimum section properties for a minimum weight of metal. The extrusion process makes this possible. The limit to which a designer may thus go in spreading out the material depends on local buckling of the individual plate elements comprising the section. If the section is made too wide and thin, premature failure will occur with local buckles forming in elements that carry compressive stress (Figure 7.1).

The first step in checking the static strength of a member is to classify the cross-section. Is it compact or slender? If it is compact, local buckling is not a problem and may be ignored. If it is slender the resistance will be reduced, with interaction occurring between local buckling of individual plate elements and overall buckling of the member as a whole.

7.1.2 Types of plate element

Two basic kinds of element are recognized, namely *internal elements* and *outstands*. An internal element is attached to the rest of the section at both

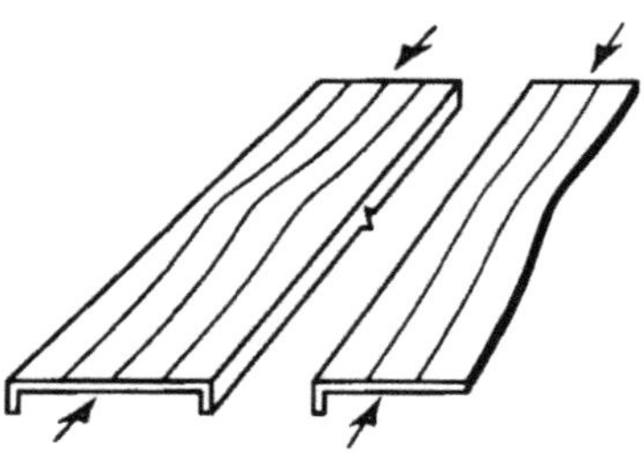

Figure 7.1 Local buckling.

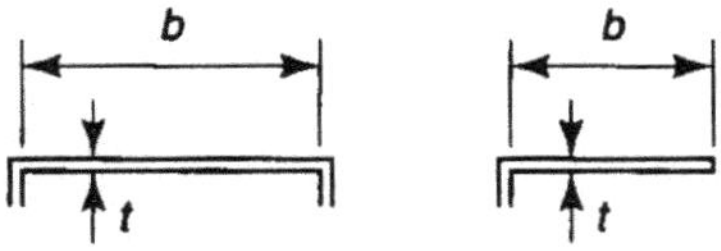

Figure 7.2 Basic element types: internal (left) and outstand (right).

its longitudinal edges, an outstand at only one (Figure 7.2). In some light-gauge steel codes, these have been confusingly referred to as 'restrained' and 'unrestrained' elements.

An element may be subjected to uniform compression (see Section 7.2), as when it forms part of a compression member or of a horizontal compression flange in a beam. Or it may be under *strain gradient* as in a beam web (see Section 7.3). Elements are usually in the form of a plain flat plate. However, it sometimes pays to improve their stability by adding a stiffener, in which case they are referred to as *reinforced* or *stiffened* elements (see Section 7.4).

7.1.3 Plate slenderness parameter

The ability of an element to resist local buckling depends on the plate slenderness parameter β which is generally taken as:

$$\beta = \frac{b}{t} \tag{7.1}$$

where b is the flat transverse width of the plate, measured to the springing of any fillet material, and t its thickness. β does not depend on the length a (in the direction of stress) because this has no effect on local buckling resistance, unless a is very small, i.e. of the same order as b.

7.1.4 Element classification (compact or slender)

In order to classify the cross-section of a member, we first classify its individual elements, excluding any that may be wholly in tension. The most adverse classification thus obtained then defines that for the member as a whole.

For any given type of element there is a critical slenderness β_s such that local buckling failure occurs just as the applied stress reaches the limiting value p_o for the material. An element having $\beta < \beta_s$ is said to be *compact*, since it is fully effective and able to reach p_o without buckling. If $\beta > \beta_s$ the element will buckle prematurely and only be partially effective, in which case it is referred to as *slender*.

(a) Compression member elements

These are simply classified as compact or slender:

$\beta \leqslant \beta_s$ Compact;
$\beta > \beta_s$ Slender.

(b) Beam elements

For these the compact classification is subdivided into fully compact and semi-compact:

$\beta \leqslant \beta_f$ Fully compact (class 2);
$\beta_f < \beta \leqslant \beta_s$ Semi-compact (class 3);
$\beta > \beta_s$ Slender (class 4).

Here β_f is a value such that an element is able, not only to attain the stress p_o, but also to accept considerably more strain while holding that stress. Thus, if all the compressed elements in a beam are fully compact, the section can achieve its full potential moment based on the plastic section modulus.

Some readers may be more familiar with the terminology used in Eurocode documents, in which sections are referred to by class numbers, as shown in brackets under (b) above. (Class 1 comprises elements of even lower β than class 2, such that 'plastic hinges' are able to operate. This is of negligible interest in aluminium.)

7.1.5 Treatment of slender elements

The buckling of slender elements is allowed for in design by taking an effective section, in the same general way as for HAZ softening. The actual cross-section of the element is replaced by an effective one which is assumed to perform at full stress, the rest of its area being regarded as ineffective. In this chapter, we advocate an *effective width* method for so doing, in preference to the *effective thickness* treatment in BS.8118. The latter is an unrealistic model of what really happens and can produce unsafe predictions in some situations.

7.2 PLAIN FLAT ELEMENTS IN UNIFORM COMPRESSION

7.2.1 Local buckling behaviour

First we consider elements under uniform compression. Figure 7.3 shows the typical variation of local buckling strength with slenderness for internal elements, plotted non-dimensionally, where: σ_m = mean stress at failure, f_o = 0.2% proof stress, E = Young's modulus, β = plate slenderness

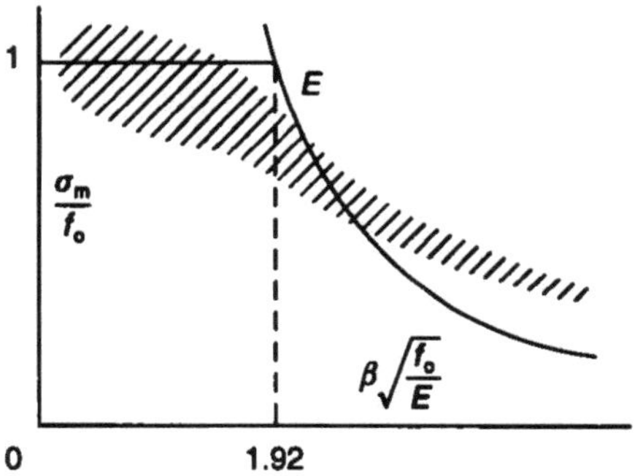

Figure 7.3 Typical relation between local buckling strength and slenderness for internal elements, plotted non-dimensionally.

(see Section 7.1.3). The pattern for outstands is similar, but with a different β-scale (about one-third). The scatter shown in the figure results from random effects including initial out-of-flatness and shape of the stress–strain curve.

Curve E indicates the stress at which buckling would begin to occur for an ideal non-welded plate, having purely elastic behaviour and zero initial out-of-flatness. For very slender plates, this stress, known as the *elastic critical stress* (σ_{cr}), is less than the mean applied stress σ_m at which the plate actually collapses. In other words, thin plates exhibit a post-buckled reserve of strength. In the USA the specific terms 'buckling' and 'crippling' are used to distinguish between σ_{cr} and σ_m. The elastic critical stress σ_{cr}, which is readily found from classical plate theory, is given by the following well-known expression:

$$\sigma_{cr} = K\left\{\frac{\pi^2}{12(1-\nu^2)}\right\}\frac{E}{\beta^2} \tag{7.2}$$

$$= \frac{0.923\,KE}{\beta^2} \text{ for aluminium}$$

in which the buckling coefficient K may be taken as follows for elements that are freely hinged along their attached edges or edge:

Internal element	$K = 4$
Outstand	$K = 0.41.$

It is found that welded plates (i.e. ones with longitudinal edge welds) perform less well than non-welded plates, with strengths tending to fall in the lower part of the scatter-band in Figure 7.3. This is due to HAZ softening, and also the effects of weld shrinkage (locked-in stress, distortion). Because of the HAZ softening, σ_m for compact welded plates (low β) fails to reach f_o, unlike the performance of non-welded ones.

Table 7.1 Classification of elements under uniform compression – limiting β-values

	Non-welded	Welded
Internal elements		
β_f	18ε	15ε
β_s	22ε	18ε
Outstands		
β_f	6ε	5ε
β_s	7ε	6ε

Note: $\varepsilon = \sqrt{250/p_o}$ where p_o is in N/mm^2.

7.2.2 Limiting values of plate slenderness

Table 7.1 lists limiting values of β_f and β_s needed for the classification of elements under uniform compression (see Section 7.1.4). These have been taken from BS.8118 and are expressed in terms of the parameter ε given by:

$$\varepsilon = \sqrt{\frac{250}{p_o}} \tag{7.3}$$

where the limiting stress p_o is measured in N/mm^2. This parameter is needed because β_f and β_s depend not only on the type of element, but also on the material properties: the stronger the metal, the more critical is the effect of buckling. Note that ε is roughly equal to unity for 6082-T6 material or equivalent.

7.2.3 Slender internal elements

For a slender non-welded internal element under uniform compressive strain, the typical form of the stress pattern at collapse is as shown by curve 1 in Figure 7.4, with the load mainly carried on the outer parts of the plate. For design purposes, we approximate to this by taking an idealized stress pattern 2, comprising equal stress blocks at either edge which operate at the full stress p_o with the material in the middle regarded as ineffective. The width of the two stress blocks is notionally adjusted to make their combined area equal to that under curve 1. For a non-welded element, therefore, the assumed effective section is as shown in diagram N with equal blocks of width b_{e1} and thickness t.

A modified diagram is needed if the element contains edge welds. First, the effective block widths must be decreased to allow for the adverse effects of weld shrinkage (locked-in stress, greater initial out-of-flatness). Secondly, an allowance must be made for the effects of HAZ softening. Diagram W shows the effective section thus obtained for a welded plate having the same β. The block widths b_{e1} are seen to be less; while in the assumed HAZ a reduced thickness of $k_z t$ is taken, where k_z is the HAZ softening factor (Section 6.4).

For design purposes the effective block width b_{e1} can be obtained from the following general expression:

$$b_{e1} = \alpha_1 \varepsilon t \tag{7.4}$$

in which α_1 is a function of β/ε and ε is as defined by equation (7.3). The value of α_1 may be calculated from the formula:

$$\alpha_1 = P_1 - \frac{Q_1}{\Delta} \tag{7.5}$$

where $\Delta = \beta/\varepsilon$ and P_1 and Q_1 are given in Table 7.2.

Figure 7.6 shows curves of α_1 plotted against Δ covering non-welded and edge-welded plates. Note that this design data, if expressed in terms of σ_m, would produce curves appropriately located (low down) in the scatter band in Figure 7.3, taking advantage of post-buckled strength at high β. It is based on the results of a parametric study by Mofflin, supported by tests [24], and also those of Little [31].

The above treatment contrasts with the effective thickness approach in BS.8118. This is a less realistic model, in which the plate is assumed to be effective over its full width, but with a reduced thickness. When applied to non-welded plates, it gives the same predictions as ours. But for welded ones it tends to be unsafe, because it makes an inadequate correction for HAZ softening, or none at all at high β.

7.2.4 Slender outstands

We now turn to outstands, again under uniform compression. For a slender non-welded outstand the stress pattern at collapse will be of the typical form shown by curve 1 in Figure 7.5, with the load mainly carried by the material at the inboard edge. The idealized pattern used in design is indicated by curve 2, with a fully effective stress block next to this edge and the tip material assumed ineffective. Our effective section is therefore as shown in figure diagram N (non-welded) or W (with an edge weld).

The effective block width b_{eo} may generally be obtained using a similar expression to that for internal elements, namely:

$$b_{eo} = \alpha_o \varepsilon t \tag{7.6}$$

where α_o is again a function of the plate slenderness and ε is given by equation (7.3). Here α_o can be calculated from:

$$\alpha_o = P_o - \frac{Q_o}{\Delta} \tag{7.7}$$

where $\Delta = \beta/\varepsilon$, and P_o and Q_o are as given in Table 7.2.

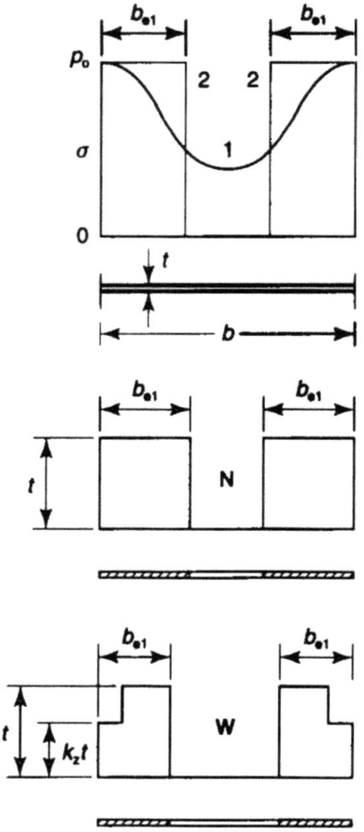

Figure 7.4 Slender internal element. Stress-pattern at failure, and assumed effective section. N = non-welded, W = with edge-welds.

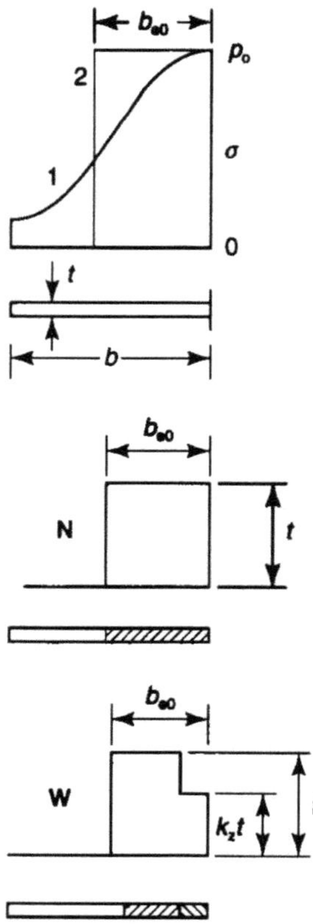

Figure 7.5 Slender outstand. Stress-pattern at failure, and assumed effective section. N = non-welded, W = welded at the connected edge.

Table 7.2 Effective section of slender elements – coefficients in the formulae for α_1 and α_o

	Non-welded	Welded
Internal elements		
P_1	16	14.5
Q_1	110	99
Outstands		
P_o	11	10
Q_o	28	24

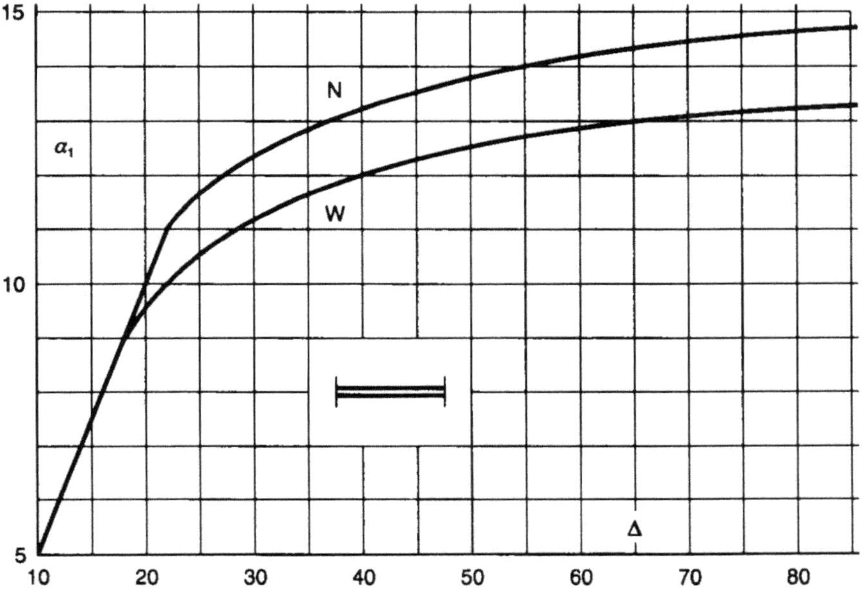

Figure 7.6 Internal elements, effective width coefficient α_1. N = non-welded, W = with edge-welds.

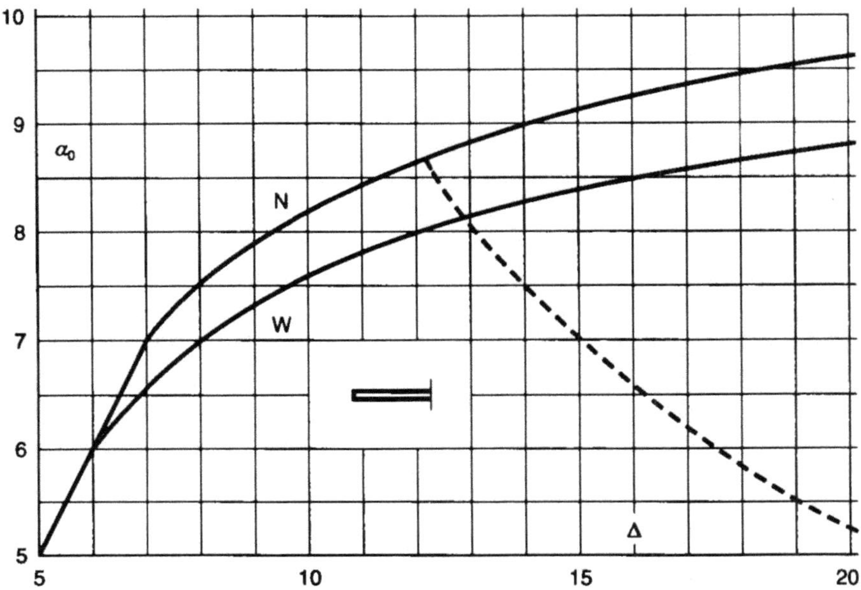

Figure 7.7 Outstands, effective width coefficient α_o. N = non-welded, W = welded at connected edge. Broken line relates to strength based on initial buckling (σ_{cr}).

Figure 7.7 shows curves of α_o plotted against Δ, covering the non-welded and edge-welded cases.

7.2.5 Very slender outstands

By a 'very slender' plate element we mean one of high β that is able to develop extra strength after the initial onset of buckling ($\sigma_m > \sigma_{cr}$) (Figure 7.3). Expressions (7.5) and (7.7) take advantage of the post-buckled reserve of strength in such elements. In the case of a very slender outstand, such an approach is sometimes unacceptable, because of the change in the stress pattern in the post-buckled state, whereby load is shed from the tip of the outstand to its root (curve 1 in Figure 7.5). The effective minor axis stiffness of an I-section or channel containing very slender flanges is thereby seriously reduced, because the flange tips become progressively less effective as buckling proceeds. Also, with the channel, there is the possibility of an effective eccentricity of loading, as the centre of resistance for the flange material moves towards the connected edge. Both effects tend to reduce the resistance of the member to overall buckling.

When necessary, the designer may allow for these effects by taking a reduced effective width for very slender outstands, based on initial buckling (σ_{cr}) rather than σ_m. This is effectively achieved by using the following expression instead of equation (7.7):

$$\alpha_o = \frac{105}{\Delta} \tag{7.8}$$

which becomes operative when:

Non-welded outstand	$\beta > 12.1\varepsilon$
Welded outstand	$\beta > 12.9\varepsilon$.

The effect of so doing is shown by the broken curve in Figure 7.7. Chapters 8 and 9 explain when it is necessary to use expression (7.8) rather than (7.7).

With beams, when considering the moment resistance of a local cross-section (Section 8.2), it is permissible to take advantage of the post-buckled strength of a very slender outstand and work to the relevant full line in figure 7.7 (equation (7.7)). But in dealing with LT buckling of such a member, it may be necessary to assume a reduced effective section based on initial buckling (σ_{cr}). Refer to Section 8.7.6.

With compression members it is again acceptable to take advantage of post-buckled strength, when studying failure at a localized cross-section (Section 9.3). And when considering overall buckling of the member as a whole, again allowance may have to be made for the loss of stiffness when the applied stress reaches σ_{cr}. Refer to Sections 9.5.4, 9.6.9.

7.3 PLAIN FLAT ELEMENTS UNDER STRAIN GRADIENT

We now consider the strain-gradient case, covering any element in a beam that is not parallel to the neutral axis, such as a web or an inclined flange element. The problem is how to apply the basic local buckling data, as established for uniform compression, to an element under strain gradient.

7.3.1 Internal elements under strain gradient, general description

First we consider internal elements, for which edges 1 and 2 may be identified as follows (Figure 7.8):

Edge 1	the more heavily compressed edge;
Edge 2	the other edge.

A parameter ψ is introduced to describe the degree of strain gradient:

$$\psi = \frac{e_2}{e_1}$$

where e_1 and e_2 are the strains arising at the two edges under simple beam theory ('plane sections remain plane'). Two cases arise:

$1 > \psi > 0$	inclined flange elements (edge 2 is in compression);
$\psi < 0$	web elements (edge 2 is in tension).

For web elements ($\psi < 0$), we use the symbol d to denote the plate width, rather than b, made up of widths d_c in compression and d_t in tension. The

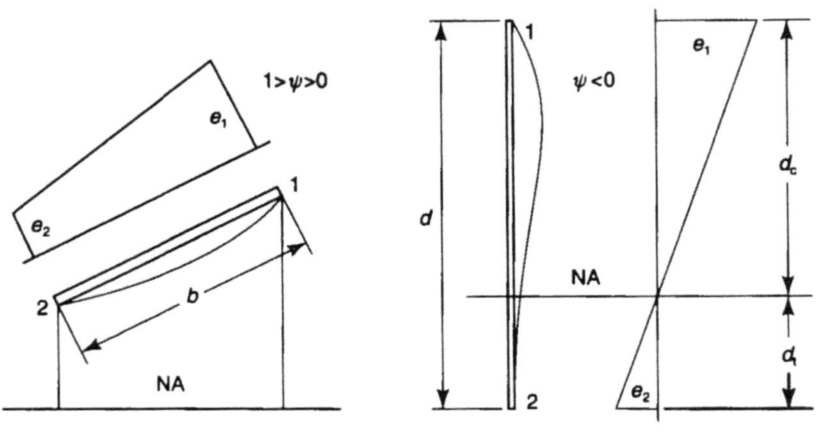

Figure 7.8 Internal elements under strain gradient. NA = neutral axis.

plate slenderness β is now taken as follows:

$$1 > \psi > 0 \qquad \beta = \frac{b}{t} \tag{7.9a}$$

$$\psi < 0 \qquad \beta = \frac{d_c}{t} \tag{7.9b}$$

The strain gradient case is always more favourable than that of uniform compression ($\psi = 1$), because the peak strain only arises at one point in the width of the element, namely at edge 1. The strain at this point at the onset of buckling is higher than it would be for a uniformly compressed element of the same β. Also the buckled shape is asymmetric, with the maximum depth of buckle occurring nearer to edge 1 than edge 2.

The elastic critical stress σ_{cr}, which now refers to the stress at edge 1, is a function of ψ. It may be calculated [25] from the standard expression (7.2) with K found from the following empirical formulae (plotted in figure 7.9), which are close enough to the exact theory:

$$1 > \psi > 0 \qquad K = 4\{1 + (1-\psi)^{1.5}\} \tag{7.10a}$$

$$\psi < 0 \qquad K = 6 + 2(1-\psi)^{-6} \tag{7.10b}$$

7.3.2 Internal elements under strain gradient, classification

The limiting plate slenderness β_f or β_s, needed for element classification, has the same meaning as before (Section 7.1.4). But the actual values in the

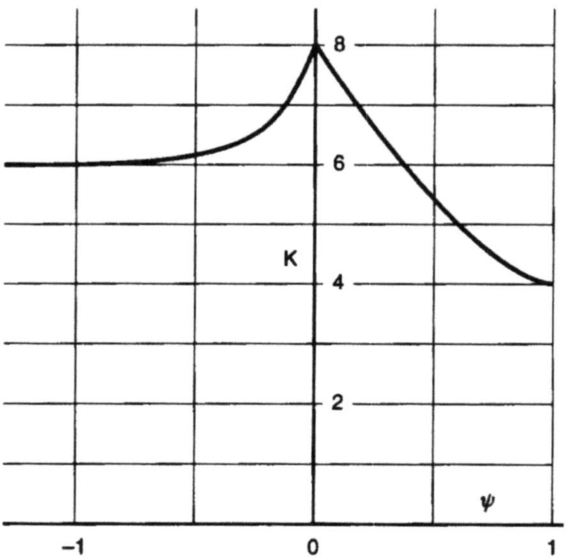

Figure 7.9 Internal elements under strain gradient, elastic buckling coefficient K.

strain-gradient case will be higher than for uniform compression. They may be read from the relevant curve in Figure 7.10, the equations to which are given in Table 7.3. Classification obtained in this manner agrees well with BS.8118.

In entering the figure, ψ should be based on the beam's plastic neutral

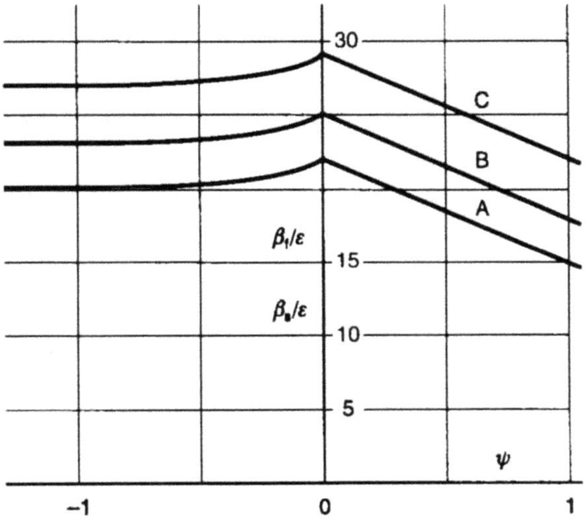

Figure 7.10 Internal elements under strain gradient, limiting values of β/ε. Non-welded: β_f = curve B; β_s = curve C. Edge-welded: β_f = curve A; β_s = curve B.

Table 7.3 Classification of plate elements under strain gradient – formulae for the limiting values of β

		Non-welded	Welded
Internal elements			
$\psi > 0$	$\beta_f =$	$(25 - 7\psi)\varepsilon$	$(22 - 7\psi)\varepsilon$
	$\beta_s =$	$(29 - 7\psi)\varepsilon$	$(25 - 7\psi)\varepsilon$
$\psi < 0$	$\beta_f =$	$\{23 + 2(1 - \psi)^{-4}\}\varepsilon$	$\{20 + 2(1 - \psi)^{-4}\}\varepsilon$
	$\beta_s =$	$\{27 + 2(1 - \psi)^{-4}\}\varepsilon$	$\{23 + 2(1 - \psi)^{-4}\}\varepsilon$
Outstands			
Case T	$\beta_f =$	6ε	5ε
	$\beta_s =$	7ε	6ε
Case R	$\beta_f =$	$6\varepsilon/g$	$5\varepsilon/g$
	$\beta_s =$	$7\varepsilon/g$	$6\varepsilon/g$

Notes. 1. Outstands: case T = peak compression at tip, case R = peak compression at root.
2. $\varepsilon = \sqrt{250/p_o}$ where p_o is in N/mm^2.
3. $g = 0.7 + 0.3\psi$ for $\psi > -1$; $= 0.8/(1 - \psi)$ for $\psi < -1$.

axis (equal area axis) for finding β_f and on the elastic axis (through the centroid) for β_s. For convenience, the neutral axis may be based on the gross section when classifying the section. But when subsequently calculating the section properties, an effective section must be taken.

7.3.3 Slender internal elements under strain gradient

For a slender element under strain-gradient it is logical to assume that the stress blocks comprising its effective section are unequal. The block at edge 2 should be made wider than that at edge 1, thus reflecting the asymmetry of the buckled form. In our method, the two block-widths are found as follows (Figure 7.11).

(a) Inclined flange element $(1 > \psi > 0)$

$$b_{e1} = \alpha_1 \varepsilon t \tag{7.11a}$$

$$b_{e2} = (1.4 - 0.4\psi) b_{e1} \tag{7.11b}$$

where α_1 is found from equation (7.5), or Figure 7.6, still taking $\Delta = \beta/\varepsilon$.

(b) Web element $(\psi < 0)$

The total element is divided into regions of compression (d_c) and tension (d_t). Block 2 is assumed to occupy all of d_t plus an amount d_{e2} of d_c as shown, and we take:

$$d_{e1} = \alpha_1 \varepsilon t \tag{7.12a}$$

$$d_{e2} = 1.4 d_{e1} \tag{7.12b}$$

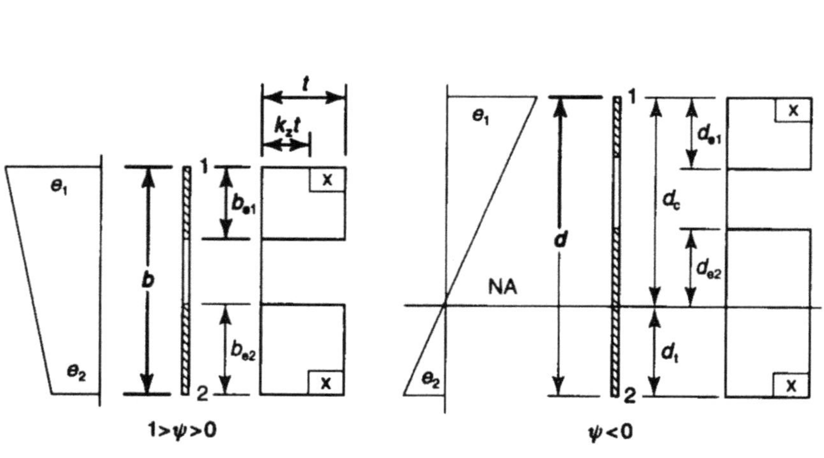

Figure 7.11 Internal elements under strain gradient, assumed effective section.

where α_1 is again found from equation (7.5), or Figure 7.6, but now putting:

$$\Delta = \frac{\beta(1-\psi)}{\varepsilon(1-1.3\psi)} \tag{7.13}$$

If an element has edge welds, a reduced thickness $k_z t$ is taken in each HAZ region in the same way as for uniform compression (i.e. the areas X in the figure are removed).

7.3.4 Outstands under strain gradient, general description

Now we turn to outstands, for which there are two possible cases of strain gradient (Figure 7.12):

Case T peak compression is at the tip;
Case R peak compression is at the root.

In either case, the more heavily compressed edge is called edge 1 as before. And again we use ψ ($=e_2/e_1$) to specify the degree of strain gradient.

The buckling of an outstand is mainly governed by what happens at the tip. Therefore, case T is only slightly more favourable than uniform compression, since the effect of ψ is merely to decrease the stress in the relatively stable material close to the supported edge. But case R is clearly more favourable, because ψ now directly affects the tip stress which is what matters.

The elastic critical stress σ_{cr} (occurring at edge 1) for a simply supported outstand under strain gradient may be found from the standard equation (7.2), taking a suitable value for K. Figure 7.13 shows how K varies with ψ for the two cases, the curves being plotted from the following empirical expressions which give a good approximation to the rigorous values:

Case T $\quad K = 0.41 \quad$ for $1 > \psi > 0.5$

$$= 0.50 - 0.18\psi \quad \text{for } \psi < 0.5 \tag{7.14a}$$

Case R

$$K = 0.41 + 1.3(1-\psi)^4 \tag{7.14b}$$

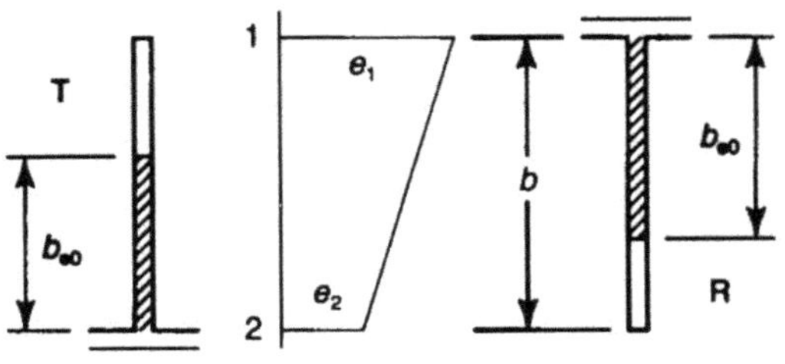

Figure 7.12 Outstands under strain-gradient. Cases T and R.

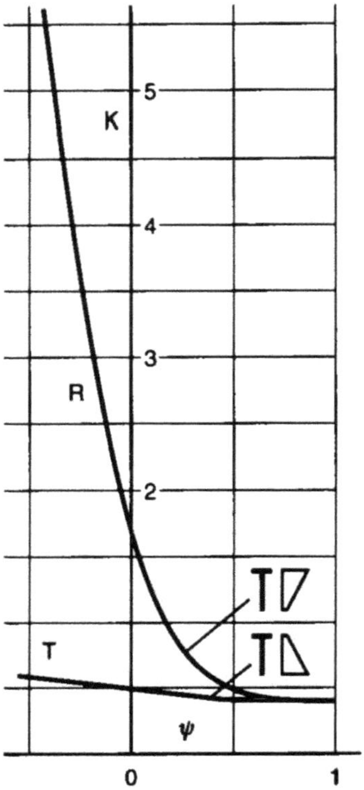

Figure 7.13 Outstands under strain-gradient, elastic buckling coefficient *K*.

7.3.5 Outstands under strain gradient, case T

For an outstand under case T loading, it is reasonable in design to ignore the slight improvement in performance as compared with uniform compression. The value of β_f or β_s needed for classification, and likewise the effective block width b_{eo} when the element is slender, may thus be taken using the data already provided for the case $\psi = 1$ (Sections 7.2.2 and 7.2.4).

7.3.6 Outstands under strain gradient, case R

With case R loading, the above approach would be unduly pessimistic. Instead we follow BS.8118 and introduce the parameter g given by:

$$1 > \psi > -1 \qquad g = 0.7 + 0.3\psi \tag{7.15a}$$

$$\psi < -1 \qquad g = \frac{0.8}{1 - \psi} \tag{7.15b}$$

Classification then consists of comparing β ($= b/t$) with modified values of

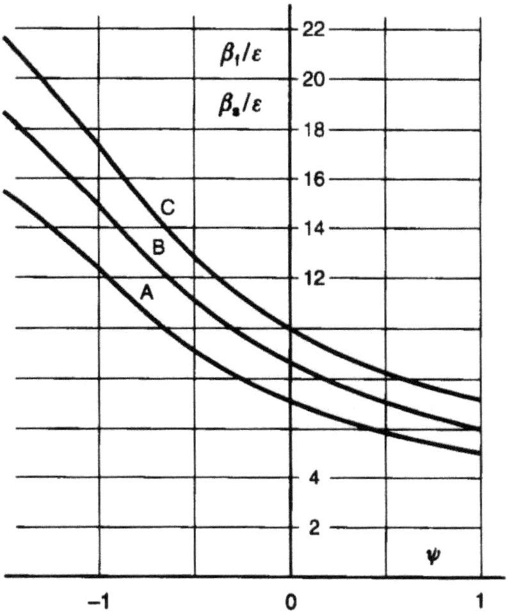

Figure 7.14 Outstands under strain-gradient (case R), limiting values of β/ε. Non-welded: β_f = curve B; β_s = curve C. Welded: β_f = curve A; β_s = curve B.

β_f and β_s obtained by dividing the uniform compression values (Table 7.1) by g. The resulting design curves of β_f and β_s plotted against ψ are given in Figure 7.14, the corresponding equations being included in Table 7.3. Again we base ψ on the plastic neutral axis of the section when finding β_f, and on the elastic one for β_s.

When the outstand is slender the effective block width b_{eo} is found from the following expression:

$$b_{eo} = \frac{\alpha_o \varepsilon t}{g} \tag{7.16}$$

where α_o is again found from equation (7.7) or Figure 7.7, but now putting:

$$\Delta = \frac{g\beta}{\varepsilon} \tag{7.17}$$

7.4 REINFORCED ELEMENTS

7.4.1 General description

The local buckling performance of a thin element can often be improved [25] by reinforcing it with a longitudinal stiffener, as shown in Figure 7.15.

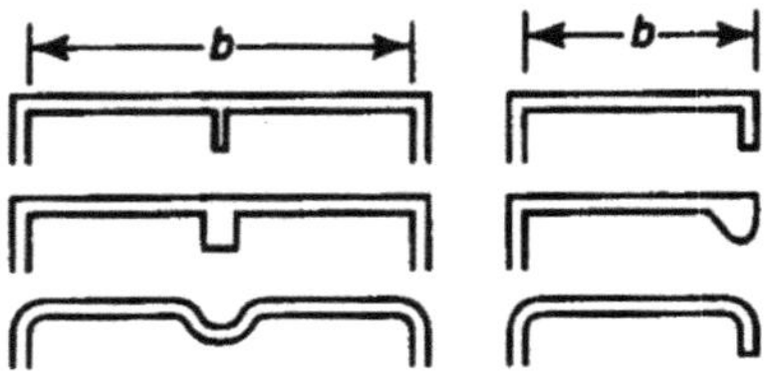

Figure 7.15 Reinforced elements.

This can take the form of an integrally extruded addition such as a rib, lip or bulb, or in a sheet metal component the reinforcement can be rolled in as a flute or lip. Very small stiffeners do nothing for plate stability, but larger ones are beneficial.

The problem is how to adapt the data provided for plain elements (Sections 7.2, 7.3) to stiffened ones. The BS.8118 method (effective thickness) is based on a study of the elastic buckling of such elements made by Bulson in the 1950s [25]. Our treatment basically follows this, but is presented in terms of effective width.

7.4.2 Limitations on stiffener geometry

The treatment assumes that a reinforced plate will buckle in the mode shown in Figure 7.16 taking the stiffener with it. For this assumption to be valid, the stiffener should be so proportioned as to be at least semi-compact when considered as an outstand element on its own (Table 7.1); otherwise it will itself buckle prematurely. If the reinforced element is to be classified as fully compact, then the stiffener must be fully compact too.

A further constraint arises in the case of beam flanges. Research by J. Rhodes at Glasgow University on light-gauge cold-rolled steel showed that a lipped outstand element forming the compression flange of a beam tends to be much less stable when the lip is outward facing. Some steel codes, therefore, require that the presence of any outward facing reinforcement should be ignored when considering the stability of such an element. A blanket restriction of this kind is unduly severe for aluminium, since the extrusion process enables one to design 'chunkier' reinforcement

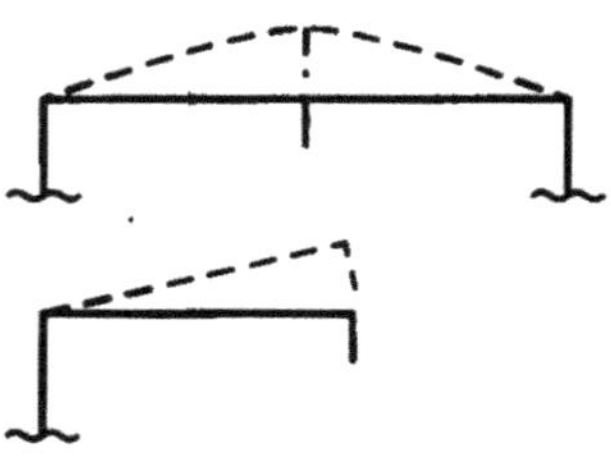

Figure 7.16 Assumed failure mode for reinforced elements.

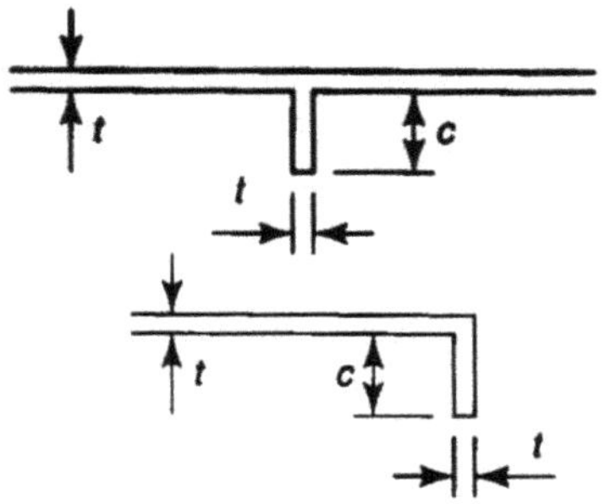

Figure 7.17 'Standard' reinforcement.

than that which is possible in a cold-rolled section. A square rib or a bulb will never become torsionally unstable, and can be safely located facing outwards. Unfortunately, no data exist as to the limiting lip geometry at which outward facing reinforcement in a beam flange begins to become unsatisfactory.

7.4.3 'Standard' reinforcement

'Standard' reinforcement comprises a plain single-sided stiffener of thickness t equal to that of the plate (Figure 7.17), the stiffener height c being measured from the near surface of the plate. The design data given below apply directly to an element stiffened thus.

For any other shape of stiffener, it is necessary to notionally replace the actual stiffener by an equivalent one of standard form, whose height c is such as to make its inertia about the mid-plane of the plate the same as that of the actual stiffener. The stability of the reinforced element is then assessed as if it had a standard stiffener with this c. Note that the equivalent standard stiffener is always taken as single-sided, even if the actual reinforcement is double-sided.

7.4.4 Location of the stiffener

(a) Internal elements under uniform compression

The stiffener is assumed to be at midwidth.

(b) Internal elements under strain gradient

We define edges 1, 2 and the parameter ψ in the same way as for the unreinforced case. It is clearly desirable to place the stiffener nearer to edge 1 than edge 2, so we assume it is located at an optimum distance j

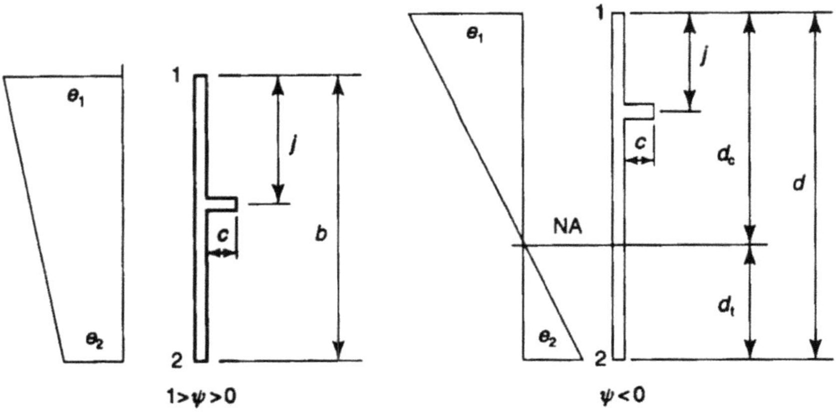

Figure 7.18 Stiffener location for internal elements under strain gradient.

from edge 1 (Figure 7.18) given by:

$$1 > \psi > 0 \qquad j = \frac{b}{2.4 - 0.4\psi} \tag{7.18a}$$

$$\psi < 0 \qquad j = \frac{d_c}{2.4} = \frac{d}{2.4(1-\psi)} \tag{7.18b}$$

(c) Outstands

The reinforcement is assumed to be at the free edge.

7.4.5 Modified slenderness parameter

The essential step in dealing with reinforced elements is to replace the usual slenderness parameter β ($=b/t$ or d_c/t) by a modified value β', defined as follows for an element having standard reinforcement:

$$\text{Reinforced internal element} \qquad \beta' = \frac{b/t}{\rho_1} \quad \text{or} \quad \frac{d_c/t}{\rho_1} \tag{7.19a}$$

$$\text{Reinforced outstand} \qquad \beta' = \frac{b/t}{\rho_0} \tag{7.19b}$$

where b is the total width or d_c the width in compression. The factors ρ (>1) are a measure of the stabilizing effect of the reinforcement, and may be read from Figure 7.19 taking c as defined in Figure 7.17. Alternatively, ρ may be calculated from an equivalent equation (valid for $c/t > 1$):

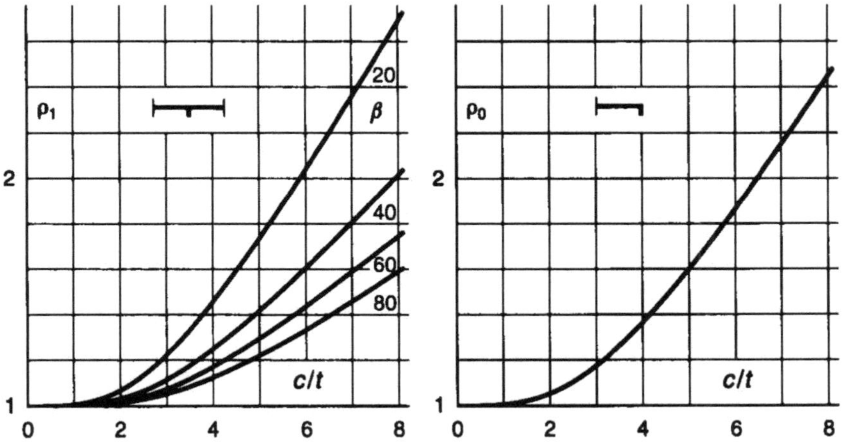

Figure 7.19 Slenderness adjustment factor for reinforced elements: internal elements ρ_1; outstands ρ_0.

$$\rho_1 = \sqrt{\left(1 + \frac{2.5(c/t - 1)^2}{\beta}\right)} \tag{7.20a}$$

$$\rho_0 = \sqrt{\{1 + 0.1(c/t - 1)^2\}} \tag{7.20b}$$

When $c < t$ the value of ρ is taken as unity, because the reinforcement then provides no benefit.

With non-standard reinforcement, the procedure is to find c for the equivalent standard stiffener and then determine ρ as above.

7.4.6 Classification

The classification (Section 7.1.4) of reinforced elements is performed simply by relating β' (instead of β) to the appropriate limiting value used for plain elements. In other words, β' is compared with β_f or β_s as given by:

Uniform compression	Table 7.1
Strain gradient, internal element	Figure 7.10
Strain gradient, outstand	Figure 7.14

For elements found to be fully or semi-compact, the full area is effective, including that of the stiffener, except where a local reduction is made at a welded edge to allow for HAZ softening.

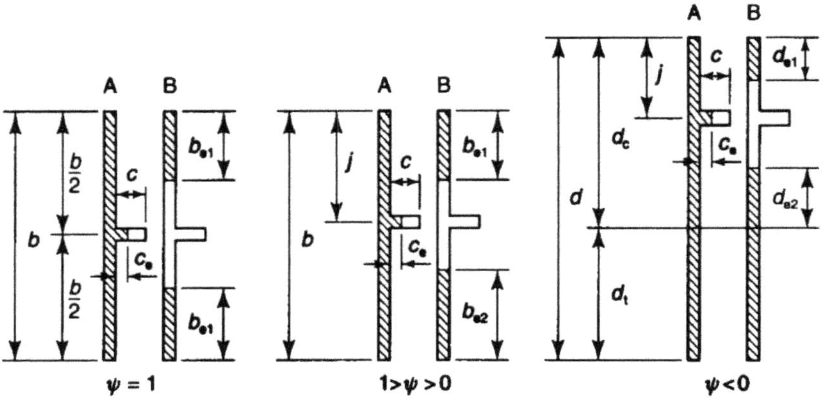

Figure 7.20 Slender reinforced internal elements, effective areas.

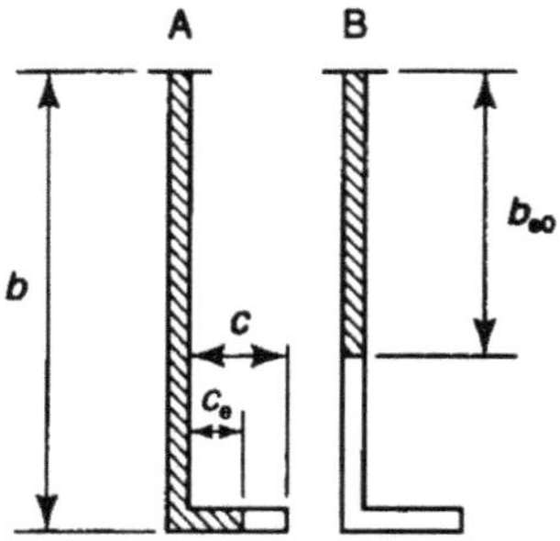

Figure 7.21 Slender reinforced outstands, effective area.

7.4.7 Slender reinforced elements

The following treatment applies to slender elements having standard reinforcement. With other forms of reinforcement, it is recommended that the load carrying capacity should be based on that of an element having equivalent standard reinforcement (Section 7.4.3).

There are two possible patterns (A, B) for the effective section of such elements, as depicted in Figures 7.20 and 7.21:

Pattern A All the plate is effective, and also a width c_e of the stiffener next to it;

Pattern B The effective section comprises two stress blocks within the plate (for an internal element) or one such stress block (for an outstand).

Pattern A applies for elements in which β' is only a bit over β_s, and pattern B for more slender ones.

Table 7.4 Slender reinforced elements, effective width formulae

		Pattern B	Pattern A	Δ
Internal elements				
$\psi = 1$		$b_{e1} = \rho_1 \alpha_1 \varepsilon t \left(1 + \frac{c}{b}\right)$	$c_e = b - 2b_{e1}$	$\Delta = \frac{\beta'}{\varepsilon}$
$1 > \psi > 0$		$b_{e1} = \rho_1 \alpha_1 \varepsilon t \left(1 + \frac{c}{b}\right)$ $b_{e2} = (1.4 - 0.4\psi) b_{e1}$	$c_e = b - (b_{e1} + b_{e2})$	$\Delta = \frac{\beta'}{\varepsilon}$
$\psi < 0$		$d_{e1} = \rho_1 \alpha_1 \varepsilon t \left(1 + \frac{c}{d_c}\right)$ $d_{e2} = 1.4 d_{e1}$ where $d_c = d/(1 - \psi)$	$c_e = d_c - (d_{e1} + d_{e2})$	$\Delta = \frac{\beta'(1 - \psi)}{\varepsilon(1 - 1.3\psi)}$
Outstands				
$\psi = 1$		$b_{e0} = \rho_0 \alpha_0 \varepsilon t \left(1 + \frac{c}{b}\right)$	$c_e = b - b_{e0}$	$\Delta = \frac{\beta'}{\varepsilon}$
$\psi < 1$				
	Case T:	Use the $\psi = 1$ formulae		
	Case R:	$b_{e0} = \frac{\rho_0 \alpha_0 \varepsilon t}{g} \left(1 + \frac{c}{b}\right)$	$c_e = b - b_{e0}$	$\Delta = \frac{g\beta'}{\varepsilon}$

Notes. 1. Refer to Figures 7.20, 7.21.
2. ψ is as defined in Section 7.3.1.
3. α_1 and α_0 are found by entering the relevant curve in Figure 7.6 or 7.7 at the value of Δ listed in final column.
4. ρ_1, ρ_0 and β' are as defined in Section 7.4.5.
5. $\varepsilon = \sqrt{250/p_0}$ where p_0 is in N/mm^2.
6. $g = 0.7 + 0.3\psi$ for $\psi > -1$, $= 0.8/(1 - \psi)$ for $\psi < -1$

Table 7.4 provides expressions which enable the stress block width or widths for pattern B to be found directly for the various cases. When the total effective width thus calculated exceeds the plate width, this means that pattern B no longer applies and pattern A becomes valid instead. The 'overflow' then defines the width of the effective part of the stiffener. In other words, c_e in pattern A is taken as the amount by which the total effective width, as calculated using the pattern B formulae, exceeds the actual plate width.

It will be seen that the pattern B formulae involve the coefficient α_1 or α_0 which may be calculated from equation (7.5) or (7.7), or read from Figure 7.6 or 7.7. In so doing, the parameter Δ should be taken as in the last column of the table.

If an element is welded along its connected edge or edges a reduced thickness of $k_z t$ is taken in the HAZ, as usual.

CHAPTER 8

Beams

8.1 GENERAL APPROACH

This chapter covers the performance of aluminium beams under static loading, for which the main requirement is strength (limit state of static strength). In order to check this there are, as in steel, four basic failure modes to consider:

1. bending-moment failure;
2. shear-force failure;
3. web crushing;
4. lateral-torsional (LT) buckling.

The resistance to bending failure (Section 8.2) must be adequate at any cross-section along the beam, and likewise for shear force (Section 8.3). When high moment and high shear act simultaneously at a cross-section, it is important to allow for their combined effect (Section 8.4). The possibility of web-crushing arises at load or reaction points, especially when there is no web stiffener fitted (Section 8.5). Lateral-torsional buckling becomes critical for deep narrow beams in which lateral supports to the compression flange are widely spaced (Section 8.7).

In checking static strength, the basic requirement is that the relevant factored resistance should not be less than the magnitude of the moment or force arising under factored loading (Section 5.1.3). The factored resistance is found by dividing the calculated resistance by the factor γ_m. The main object of this chapter is to provide means for obtaining the calculated resistance corresponding to the various possible modes of static failure. The suffix c is used to indicate 'calculated resistance'.

When a member is required to carry simultaneous bending and axial load, it is obviously necessary to allow for interaction of the two effects. This is treated separately at the end of Chapter 9.

For aluminium beams, it is also important to consider deflection (serviceability limit state) bearing in mind the metal's low modulus. This is a matter of ensuring that the elastic deflection under nominal loading (unfactored service loading) does not exceed the permitted value (Section 8.8).

8.2 MOMENT RESISTANCE OF THE CROSS-SECTION

8.2.1 Moment–curvature relation

In this section we consider resistance to bending moment. The object is to determine the calculated moment resistance M_c of the cross-section, i.e. its failure moment.

Figure 8.1 compares the relation between bending moment and curvature for a steel universal beam of fully-compact section and an extruded aluminium beam of the same section. It is assumed that the limiting stress p_o is the same for each, this being equal to the yield stress and the 0.2% proof stress respectively for the two materials. Typically, the diagram might be looked on as comparing mild steel and 6082-T6 aluminium.

Moment levels Zp_o and Sp_o are marked on the diagram, where Z and S are the elastic and plastic section moduli respectively. Both curves begin to deviate from linear at a moment below Zp_o. This happens in steel, despite the well-defined yield point of the material, because of the severe residual stresses that are locked into all steel profiles. For the aluminium, it is mainly a function of the rounded knee on the stress–strain curve.

The problem in aluminium is how to decide on an appropriate value for the limiting moment, i.e. the calculated moment resistance (M_c). This is the level of moment corresponding to 'failure' of the cross-section, at which severe plastic deformation is deemed to occur. For a steel beam, there is an obvious level at which to take this, namely at the 'fully plastic moment' Sp_o where the curve temporarily flattens out. In aluminium, although there is no such plateau, it is convenient to take the same value as for steel, namely $M_c = Sp_o$. That is what BS.8118 does and we follow suit in this book (for fully compact sections). F.M. Mazzolani has proposed a more sophisticated treatment [26].

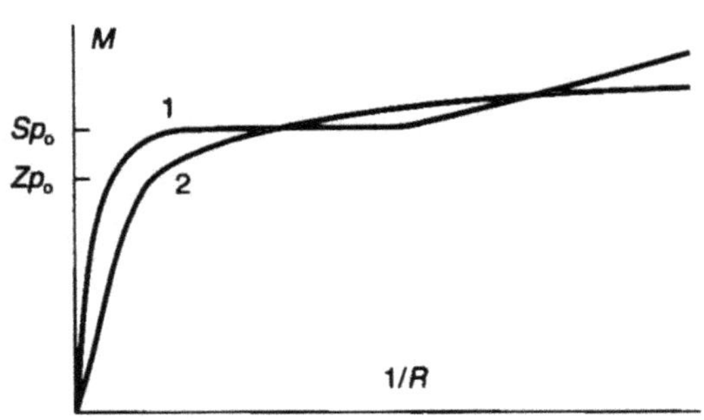

Figure 8.1 Comparison of the curves relating bending moment M and curvature $1/R$ for steel (1) and aluminium (2) beams of the same section and yield/proof stress.

8.2.2 Section classification

The discussion in Section 8.2.1 relates to beams that are thick enough for local buckling not to be a factor. We refer to these as *fully compact*. For thinner beams (*semi-compact* or *slender*) premature failure occurs due to local buckling of plate elements within the section, causing a decrease in M_c below the ideal value Sp_o.

The first step in determining M_c is to classify the section in terms of its susceptibility to local buckling (see Chapter 7). To do this the slenderness β (= b/t or d_c/t) must be calculated for any individual plate element of the section that is wholly or partly in compression, i.e. for elements forming the compression flange and the web. Each such element is then classified as fully compact, semi-compact or slender by comparing its β with the limiting values β_f and β_s as explained in Section 7.1.4.

The least favourable element classification then dictates the classification of the section as a whole. Thus, in order for the section to be classed as fully compact, all the compressed or partly compressed elements within it must themselves be fully compact. If a section contains just one slender element, then the overall section must be treated as slender. The classification is unaffected by the presence of any HAZ material.

In classifying an element under strain gradient, the parameter ψ (Section 7.3) should relate to the neutral axis position for the gross section. In checking whether the section is fully compact, this should be the plastic (equal area) neutral axis, while for a semi-compact check it should be the elastic neutral axis (through the centroid).

8.2.3 Uniaxial moment, basic formulae

First we consider cases when the moment is applied either about an axis of symmetry, or in the plane of such an axis, known as *symmetric bending* (Figure 8.2). For these the calculated moment resistance M_c of the section is normally taken as follows:

Fully compact section $\qquad M_c = Sp_o \qquad (8.1)$

Semi-compact or slender section $\qquad M_c = Zp_o \qquad (8.2)$

where p_o = limiting stress for the material (Section 5.2), S = plastic section

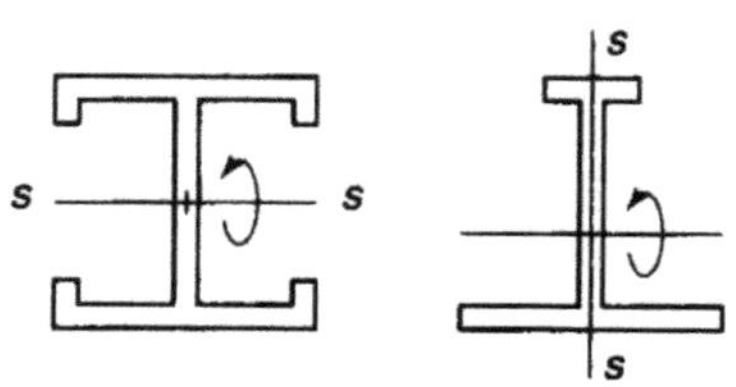

Figure 8.2 Symmetric bending.

modulus, and Z = elastic section modulus. By using these formulae we are effectively assuming idealized stress patterns (fully plastic or elastic), based on an elastic–perfectly plastic steel-type stress–strain curve. The effect, that the actual rounded nature of the stress–strain curve has on the moment capacity, is considered by Mazzolani in his book *Aluminium Alloy Structures* [26].

For a non-slender section, non-welded and without holes, the modulus S or Z is based on the actual *gross* cross-section. Otherwise, it should be found using the *effective* section (Section 8.2.4). Chapter 10 gives guidance on the calculation of these section properties.

8.2.4 Effective section

The moment resistance of the cross-section must when necessary be based on an effective section rather than the gross section, so as to allow for HAZ softening at welds, local buckling of slender plate elements or the presence of holes:

1. *HAZ softening*. In order to allow for the softening at a welded joint, we assume that there is a uniformly weakened zone of nominal area A_z beyond which full parent properties apply. One method is to take a reduced thickness $k_z t$ in this zone and calculate the modulus accordingly, the appropriate value for the softening factor k_z being k_{z2}. Alternatively the designer can assume that there is a 'lost area' of $A_z (1 - k_z)$ at each HAZ, and make a suitable deduction from the value obtained for the modulus with HAZ softening ignored. Refer to Section 6.6.1.
2. *Local buckling*. For a slender plate element, it is assumed that there is a block of effective area adjacent to each connected edge, the rest of the element being ineffective (Chapter 7). For a very slender outstand element (Section 7.2.5), as might occur in an I-beam compression flange, it is permissible to take advantage of post-buckled strength when finding the moment resistance of the cross-section, rather than use the reduced effective width corresponding to initial buckling.
3. *Holes*. The presence of a small hole is normally allowed for by removing an area dt from the section, where d is the hole diameter, although filled holes on the compression side can be ignored. However, for a hole in the HAZ, the deduction need be only $k_z dt$, while for one in the ineffective region of a slender element none is required.

In dealing with the local buckling of slender elements under strain gradient, as in webs, it is normal for convenience to base the parameter ψ (Section 7.3) on the neutral axis for the gross section, and obtain the widths of the effective stress-blocks accordingly. However, in then going on to calculate the effective section properties (I, Z), it is essential to use the neutral axis of the effective section, which will be at a slightly different position from that for the gross section. It would in theory be more accurate to adopt an iterative procedure, with ψ adjusted according to the centroid position of the effective section, although in practice nobody does so.

For a beam containing welded transverse stiffeners, two alternative effective sections should be considered. One is taken mid-way between stiffeners, allowing for buckling. The other is taken at the stiffener position with HAZ softening allowed for, but buckling ignored.

8.2.5 Hybrid sections

It is possible for aluminium beams to be fabricated from components of differing strength, as for example when 6082-T6 flanges are welded to a 5154A-H24 web. The safe but pessimistic procedure for such a hybrid section is to base design on the lowest value of p_o within the section. Alternatively, one can classify the section taking the true value of p_o for each element, and then proceed as follows:

1. *Fully compact section.* Conventional plastic bending theory is used, M_c being based on an idealized stress pattern, in which due account is taken of the differing values of p_o.
2. *Other sections.* Alternative values for M_c are found, of which the lower is then taken. One value is obtained using equation (8.2) based on p_o for the extreme fibre material. The other is found as follows:

$$M_c = \left(\frac{I}{y}\right) p_o \tag{8.3}$$

where I = second moment of area of effective section, y = distance from neutral axis thereof to the edge of the web, or to another critical point in a weaker element of the section, p_o = limiting stress for the web or other weaker element considered.

8.2.6 Use of interpolation for semi-compact sections

Consider the typical section shown in Figure 8.3, when the critical element X is just semi-compact ($\beta = \beta_s$). In such a case, equation (8.2) gives a good

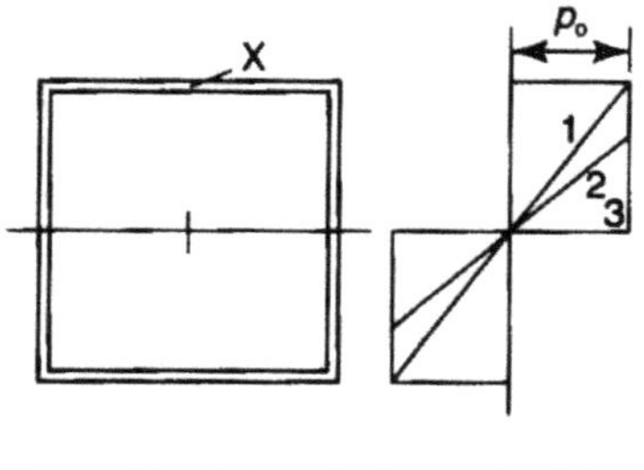

Figure 8.3 Interpolation method for semi-compact beam, idealized stress-patterns.

estimate for the resistance M_c of the section, since element X can just reach a stress p_o before it buckles. Line 1 in the figure, on which equation (8.2) is based, approximately represents the stress pattern for this case when failure is imminent. (We say 'approximately' because it ignores the rounded knee on the stress–strain curve.)

Now consider semi-compact sections generally, again taking the type of beam in Figure 8.3 as an example. For these, the critical element X will have a β-value somewhere between β_f and β_s, and some degree of plastic straining can therefore take place before failure occurs. This leads to an elasto-plastic stress pattern at failure such as line 2 in the figure, corresponding to a bending moment in excess of the value Zp_o. In the extreme case when element X is almost fully compact (β only just greater than β_f) the stress pattern at failure approaches line 3, corresponding to an ultimate moment equal to the fully compact value Sp_o which can be as much as 15% above that based on Z. It is thus seen that the use of the value Zp_o tends to underestimate M_c, increasingly so as β for the critical element approaches β_f.

It is therefore suggested that interpolation should be used for semi-compact sections, with M_c found as follows:

$$M_c = Zp_o + \frac{\beta_s - \beta}{\beta_s - \beta_f}(Sp_o - Zp_o) \qquad (8.4)$$

This expression, in which the β's refer to the critical element, will produce higher values of M_c closer to the true behaviour.

8.2.7 Semi-compact section with tongue plates

Figure 8.4 shows a section with tongue plates, in which the d/t of the web (between tongues) is such as to make it semi-compact when classified in

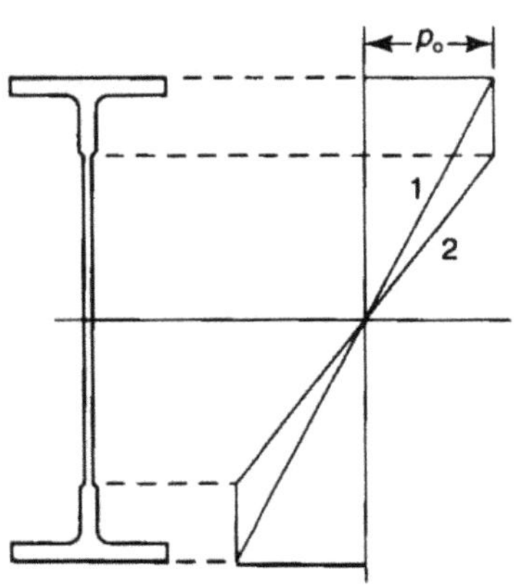

Figure 8.4 Elastic–plastic method for beam with tongue plates.

the usual manner. Line 1 in the figure indicates the (idealized) stress pattern on which M_c would be normally based, corresponding to expression (8.2). Clearly this fails to utilise the full capacity of the web, since the stress at its top edge is well below the value p_o it can attain before buckling. Equation (8.2) therefore underestimates M_c.

An improved result may be obtained by employing an elasto-plastic treatment, in which a more favourable stress pattern is assumed with yield penetrating to the top of the web (line 2). The moment calculation is then based on line 2 instead of line 1. If the web is only just semi-compact, M_c is taken equal to the value M_{c2} calculated directly from line 2, while, in the general semi-compact case, it is obtained by interpolation using equation (8.4), with the quantity Zp_o replaced by M_{c2}.

When operating this method we allow for HAZ softening by using the gross section and reducing the stress in any HAZ region to $k_{z2}p_o$.

8.2.8 Local buckling in an understressed compression flange

Figure 8.5 shows a type of section in which the distance y_c from the neutral axis to a slender compression flange X is less than the distance y_t to the tension face. For such sections, local buckling in the compression flange becomes less critical because it is 'understressed', and the normal method of calculation will produce an oversafe estimate of M_c. An improved result can be obtained by replacing the parameter ε ($= \sqrt{(250/p_o)}$) in the local buckling calculations by a modified value ε' given by:

$$\varepsilon' = \sqrt{\frac{250y_t}{p_o y_c}} \tag{8.5}$$

This may make it possible to re-classify the section as semi-compact, or, if it is still slender, a more favourable effective section will result. In either case, M_c is increased.

Note, however, that such a device should not be employed as a means for upgrading a section from semi-compact to fully compact.

8.2.9 Biaxial moment

We now turn to the case of asymmetric bending, when the applied moment has components about both the principal axes of the section.

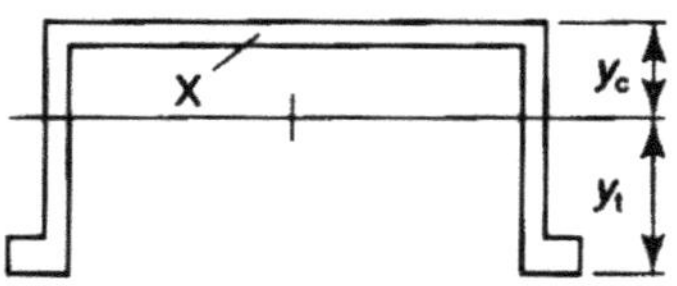

Figure 8.5 Understressed compression flange (X).

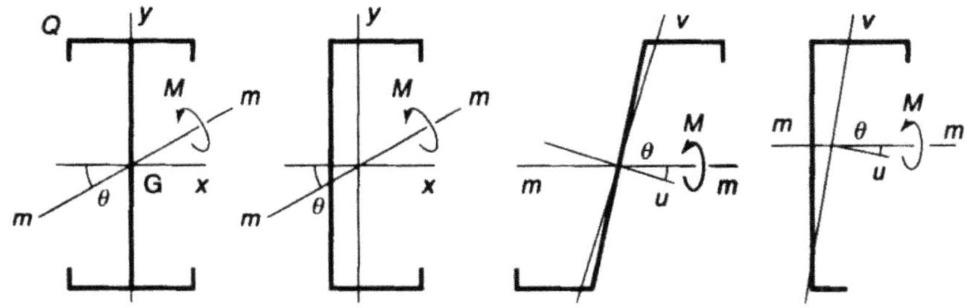

Figure 8.6 Asymmetric bending cases.

Examples of this are shown in Figure 8.6:

(a, b) bisymmetric or monosymmetric section with inclined moment;
(c) skew-symmetric section;
(d) asymmetric section.

For (a) and (b), the essential difference from symmetric bending is that the neutral axis (axis of zero stress) no longer coincides with the axis *mm* of the applied moment *M*. The same applies to (c) and (d), unless *mm* happens to coincide with a principal axis of the section. Also, for any given inclination of *mm* the plastic and elastic neutral axes will be orientated differently.

In classifying the section, the parameter ψ for any element under strain gradient should be based on the appropriate neutral axis, which properly relates to the inclination of *mm*. This should be the plastic neutral axis for the fully-compact check (Section 10.2.2), or the elastic one for the semi-compact check (see 2 below).

Having classified the cross-section, a simple procedure is then to use an interaction formula, such as that given in BS.8118, which for bisymmetric and monosymmetric sections may be written:

$$\frac{M \cos \theta}{M_{cx}} + \frac{M \sin \theta}{M_{cy}} < \frac{1}{\gamma_m} \tag{8.6}$$

where: M = moment arising under factored loading,
θ = inclination of *mm* (figure 8.6),
M_{cx}, M_{cy} = moment resistance for bending about G*x* or G*y*,
γ_m = material factor (see 5.1.3).

This expression may also be used for skew-symmetric and asymmetric profiles, changing *x*, *y* to *u*, *v*. It gives sensible results when applied to semi-compact and slender sections of conventional form (I-section, channel), but can be pessimistic if the profile is non-standard. In order to achieve better economy in such cases, the designer may, if desired, proceed as follows.

1. *Fully compact sections*. The limiting value of *M* is found using expression

(8.1) with the plastic modulus S replaced by a value S_m which is a function of the inclination of *mm* (angle θ). Refer to Section 10.2.2.

2. *Semi-compact and slender sections.* The applied moment M is resolved into components $M\cos\theta$ and $M\sin\theta$ about the principal axes, the effects of which are superposed elastically. A critical point Q is chosen and the section is adequate if at this position (Figure 8.6(a))

$$\left| -\frac{yM\cos\theta}{I_{xx}} + \frac{xM\sin\theta}{I_{yy}} \right| < \frac{p_o}{\gamma_m} \tag{8.7}$$

where x, y are the coordinates of Q, the I's are for the effective section, and the left-hand side is taken positive. The inclination ϕ of the neutral axis *nn* (anti-clockwise from Gx) is given by:

$$\tan\phi = \frac{I_{xx}}{I_{yy}}\tan\theta \tag{8.8}$$

The same expressions are valid for skew-symmetric and asymmetric shapes, if x, y are changed to u, v. Sometimes the critical point Q is not obvious, in which case alternative calculations must be made for possible locations thereof and the worst result taken. It is obviously important to take account of the signs of the stresses for flexure about the two axes.

Figure 8.7 gives a comparison between predictions made with the simple British Standard rule (expression (8.6)) and those obtained using the more accurate treatments given in 1 and 2 above. The figure relates to a particular form of extruded shape and shows how the limiting M varies

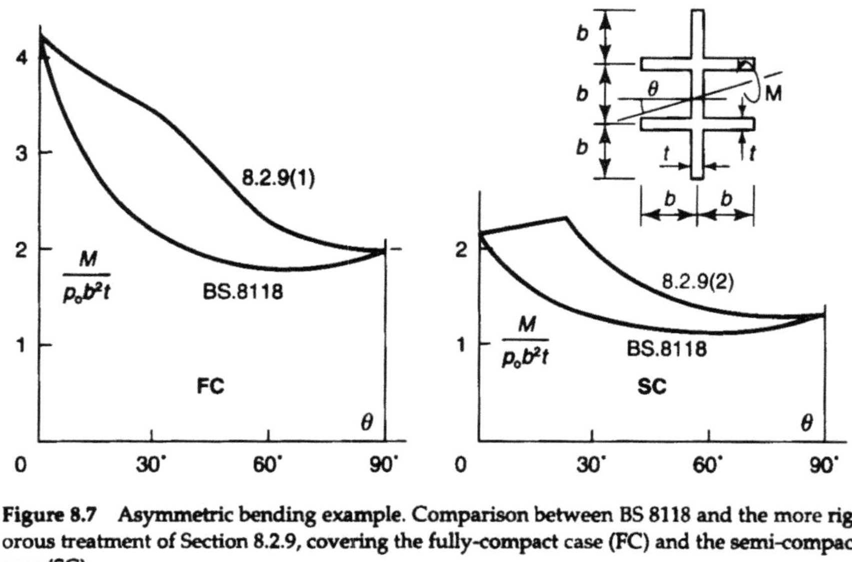

Figure 8.7 Asymmetric bending example. Comparison between BS 8118 and the more rigorous treatment of Section 8.2.9, covering the fully-compact case (FC) and the semi-compact case (SC).

with θ. It is seen that the predicted value based on expression (8.6) can be as much as 36% too low for the fully compact case and 39% too low for semi-compact.

8.3 SHEAR FORCE RESISTANCE

8.3.1 Necessary checks

We now consider the resistance of the section to shear force, for which two types of failure must be considered: (a) yielding in shear; and (b) shear buckling of the web. Procedures are presented for determining the calculated shear force resistance V_c corresponding to each of these cases.

The resistance of a thin web to shear buckling can be improved by fitting transverse stiffeners, unlike the moment resistance. This makes it difficult to provide a general rule for classifying shear webs as compact or slender. However, for simple I, channel and box-section beams, having unwelded webs of uniform thickness, it will be found that shear buckling is never critical when d/t is less than $750/\sqrt{p_o}$ where p_o is in N/mm^2. Stiffened shear webs can be designed to be non-buckling at a higher d/t than this.

8.3.2 Shear yielding of webs, method 1

Structural sections susceptible to shear failure typically contain thin vertical webs (internal elements) to carry the shear force. Alternative methods 1 and 2 are offered for obtaining the calculated resistance V_c of these, based on yield. In method 1, which is the simpler, V_c is found from the following expression:

$$V_c = 0.8Dt_1p_v \tag{8.9}$$

where D = overall depth of section, t_1 = critical thickness, and p_v = limiting stress for the material in shear (Section 5.2). In effect, we are assuming that the shear force is being carried by a thin vertical rectangle of depth D and thickness t_1 with an 80% efficiency.

For an unwelded web, t_1 is simply taken as the web thickness t or the sum of the web thicknesses in a multi-web section. If the thickness varies down the depth of the web, t_1 is the minimum thickness. When there is welding on the web, $t_1 = k_{z1}t$ where k_{z1} is the HAZ softening factor (Section 6.4).

8.3.3 Shear yielding of webs, method 2

The alternative method 2 produces a more realistic estimate of V_c which is generally higher than that given by method 1. It considers two possible

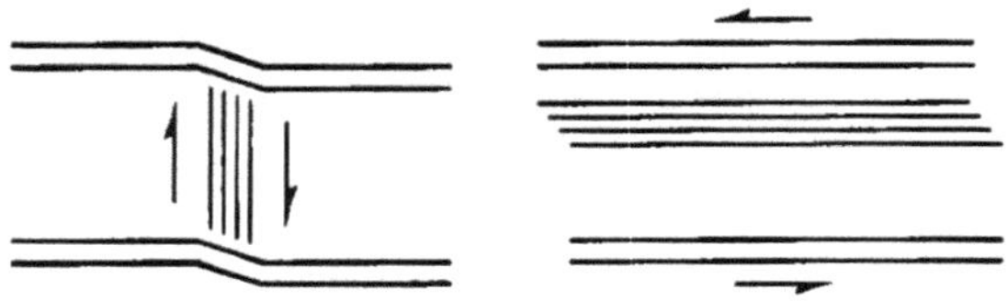

Figure 8.8 Transverse and longitudinal yield in a shear web.

patterns of yielding, corresponding to the two kinds of complementary stress that act in a web, namely transverse shear and longitudinal shear. Figure 8.8 depicts the two patterns. Transverse yielding would typically govern for a web containing a full depth vertical weld; while longitudinal yield might be critical along the line of a web-to-flange weld. (Method 1 automatically covered both patterns, conservatively.)

Method 2 employs expressions (8.10) and (8.11) for obtaining V_c, the lower value being taken. Equation (8.10) relates to yielding on a specific cross-section, and equation (8.11) to yielding on a longitudinal line at a specific distance y_v from the neutral axis.

$$\text{Transverse yield} \qquad V_c = A_w p_v \tag{8.10}$$

$$\text{Longitudinal yield} \qquad V_c = \frac{KIt_2 p_v}{(A\bar{y})} \tag{8.11}$$

where: A_w = effective section area of web,
I = inertia of the section,
t_2 = relevant thickness,
$(A\bar{y})$ = first moment of 'excluded area',
$K = 1.1$ (but see discussion below),
p_v = limiting stress for the material in shear.

A_w is taken as the effective area of all vertical material up to the face of each flange, as shown in Figure 8.9(a). Tongue plates are included but any horizontal fins are not. An effective thickness equal to $k_{z1}t$ must be assumed for any HAZ material in the web cross-section considered, where t is the actual thickness.

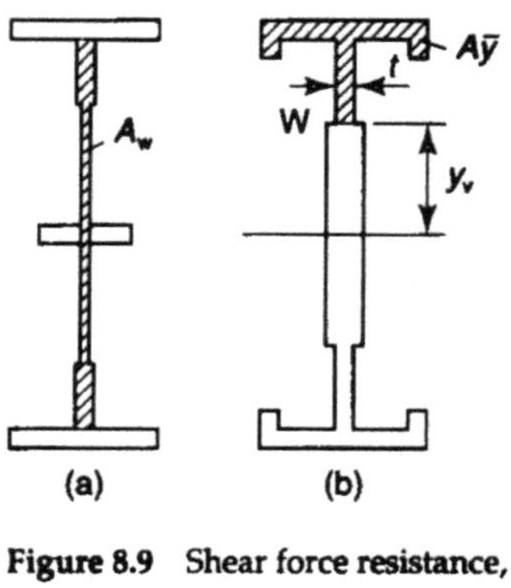

Figure 8.9 Shear force resistance, method 2. Definitions.

$(A\bar{y})$ and t_2 relate to the given point W in the web at which the longitudinal yield is being checked. Typically this point, defined by y_v, will be located in one of the following critical positions:

1. at the neutral axis ($y_v = 0$);
2. at the inmost point of a region of actual reduced thickness;
3. at the inmost point of a region of HAZ softening (reduced *effective* thickness).

In cases 1 and 2 we put t_2 equal to the actual web thickness t at the point considered, while in case 3 we put it equal to $k_{z1}t$. In a multi-web beam, t_2 should be summed for all the webs. $(A\bar{y})$ is the first moment of area about the neutral axis of all the material in the cross-section lying beyond point W, shown shaded in Figure 8.9(b).

I and $(A\bar{y})$ may for convenience both be based on the gross section. Alternatively they may both be based on the effective section as used in checking the moment.

The factor K in equation (8.11) needs consideration. If it were put equal to 1.0, this would imply that the member fails as soon as the longitudinal shear stress in the web reaches yield at the critical point, which is a pessimistic assumption. It would be more realistic to permit the area of yielding to spread slightly, and thereby allow a higher V_c. We suggest that the designer should achieve this by putting $K = 1.1$ instead of 1.0. For an ordinary unwelded I-beam, this value leads to an answer roughly equal to the method 1 value, which corresponds to the typical treatment in steel codes.

The above procedure for obtaining V_c deviates from that given in BS.8118, which appears to contain inconsistencies and shortcomings. The main point at issue concerns unwelded webs, for which the British Standard requires only the transverse check to be made. When such a web contains a region of reduced thickness, such as thin web material located between tongue plates, longitudinal yielding becomes critical. But BS.8118 calls for the longitudinal check only if there are longitudinal welds present, i.e. only for position 3. Our treatment requires it to be made for position 2 also. For an extruded I-section (unwelded) with tongue plates, it is possible for BS.8118 to over-estimate V_c by 40%, as compared with the procedure advocated above.

In applying method 2 to a hybrid section, the procedure may be modified in the following manner:

1. *Transverse check*. Equation (8.10) is used with p_v taken as that for the weaker material. Alternatively, the designer may take V_c equal to the summation of $A_e p_v$ for the various parts of the web, where A_e is the effective area of each and p_v the relevant material stress.
2. *Longitudinal check*. Equation (8.11) should be used with p_v put equal to the relevant value at the point W.

8.3.4 Shear resistance of bars and outstands

The shear force resistance V_c of a rectangular bar may be estimated using equation (8.9), with t_1 put equal to the section width. An equivalent expression may be employed for round bar or tube, replacing Dt_1 by the section area A.

When the shear force is carried by a vertical outstand element (or outstands), as in a T-bar or a channel loaded about its weak axis, V_c may be generally found by treating the outstand as a rectangular bar (as above). However, if the outstand is too slender in cross-section, its shear force resistance will be reduced by buckling. In the absence of hard data, a possible approach is to assume that the full value of V_c (based on yield) can be attained, provided the depth/thickness ratio does not exceed the limiting value β_s based on uniform compression (Table 7.1). If this value is exceeded, V_c is determined as if the extra material were not there.

8.3.5 Web buckling, simple method

The treatments given below apply to thin internal elements used as shear webs, as in an I-beam. A safe way to check for possible buckling failure in these is simply to base V_c on the initial buckling stress, and to neglect any post-buckled reserve of strength. For a plain web of uniform thickness (Figure 8.10(a)) this gives a value as follows:

$$V_c = dtp_{v1} \tag{8.12}$$

where d = depth of web between flanges, t = web thickness, p_{v1} = initial buckling stress in N/mm^2.

For an unstiffened web p_{v1} may be found from the following expression, which assumes simple support along the top and bottom edges:

$$p_{v1} = \frac{4.9E}{(d/t)^2} \tag{8.13}$$

where E is the modulus of elasticity (= 70 kN/mm^2).

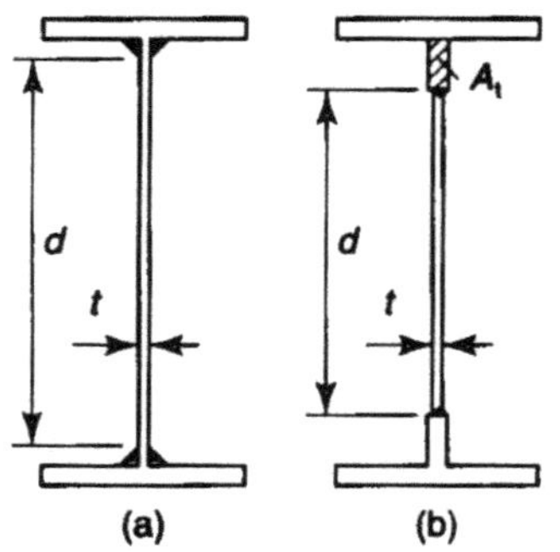

Figure 8.10 Shear buckling, web dimensions.

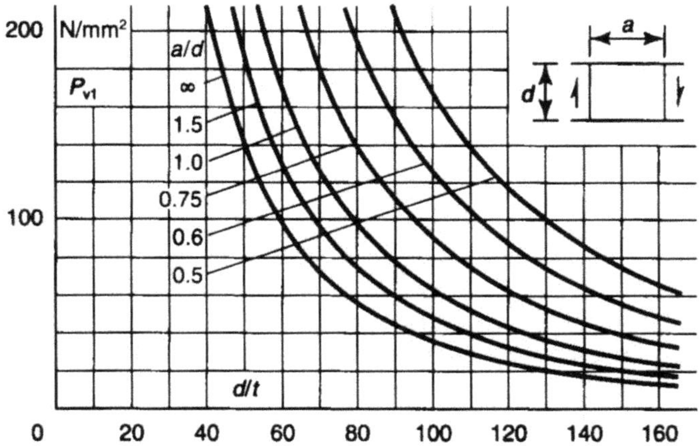

Figure 8.11 Shear buckling stress for stiffened webs (without tension-field action).

When transverse stiffeners are fitted, an improved value of p_{v1} may be read from Figure 8.11, in which a is the stiffener spacing. Alternatively the designer may use the equations on which the figure is based:

$$a > d \qquad p_{v1} = \frac{0.92E\{5.35 + 4(d/a)^2\}}{(d/t)^2} \tag{8.14a}$$

$$d > a \qquad p_{v1} = \frac{0.92E\{5.35(d/a)^2 + 4\}}{(d/t)^2} \tag{8.14b}$$

For a web fitted with tongue plate or plates (Figure 8.10(b)), it is necessary to sum the web and tongue contributions. This may be done as follows, provided the tongue plates are properly designed (Section 8.6.2):

$$V_c = V_{cw} + V_{ct} \tag{8.15}$$

where: V_{cw} = value given by (8.12) taking d as the depth between tongues,
$V_{ct} = \sum A_t p_{vt}$
$\sum A_t$ = total effective area of tongue plates,
p_{vt} = limiting stress in shear for tongue material.

In a multi-web beam, V_c may be taken as the sum of the values found for the individual webs, using expression (8.12) or (8.15) as appropriate.

8.3.6 Web buckling, tension-field action

Very slender stiffened webs can often accept a great increase in shear force above the initial buckling load before they finally fail. This results from the development of a diagonal tension field as the buckles develop (Figure 8.12). The designer can take advantage of this by using the treatment below, based on the 'Cardiff model' of Rockey and Evans [15], which

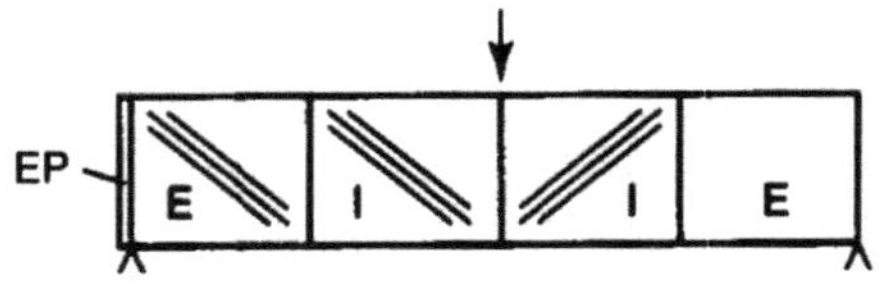

Figure 8.12 Tension-field action.

is valid when $2.5 > a/d > 0.5$. However, if the presence of visible buckles at working load is unacceptable, V_c must be found in the usual way as in Section 8.3.5, with tension-field action ignored.

In any given panel between transverse stiffeners, it is only possible for the tension field to develop if it is properly anchored at either end, i.e. if there is something for it to pull against. In an internal panel, such as I in the figure, anchorage is automatically provided by the adjacent panels. But in an end panel (E) there is generally nothing substantial for a tension field to pull on and the tension field cannot develop. Only if a proper 'end-post' (EP) is provided, designed as in Section 8.6.4, can effective tension-field action in an end-panel be assumed.

The value of V_c based on tension-field action is found as follows for a plain web (Figure 8.10(a)):

$$V_c = dt\{p_{v1} + k(v_2 + mv_3)p_v\} \tag{8.16}$$

where d, t = web depth and thickness; p_v = limiting stress in shear; p_{v1} = initial buckling stress (Section 8.3.5); m = lesser of m_1 and m_2; $k = k_{z1}$ for welded webs or 1.0 for other webs;

$$v_2 = \sqrt{3}\,Q \sin^2\theta(\cot\theta - a/d) \qquad m_1 = \sqrt{Q}\,(a/d)\sin\theta$$

$$v_3 = 2\sqrt{3Q}\sin\theta \qquad m_2 = \sqrt{4S_f/d^2t}$$

$$Q = \{1 - v_1^2(1 - 0.75\sin^2 2\theta)\}^{1/2} - (\sqrt{3}/2)v_1 \sin 2\theta$$

$$v_1 = p_{v1}/150 \qquad \theta = (2/3)\tan^{-1}(d/a)$$

S_f is plastic modulus of effective flange section about its own horizontal equal area axis, taken as the lower value when the flanges differ. In a hybrid beam the expression for m_2 should be multiplied by the square-root of the ratio of p_o for the flange material to p_o for the web.

For a web with properly designed tongue-plates V_c may be found using equation (8.15), with V_{cw} now taken as the value of V_c that would be obtained from expression (8.16) putting d equal to the web plate depth between tongues. When determining S_f it is permissible to include the tongue plate as an integral part of the flange.

In a multi-web beam, V_c is again taken as the sum of the values for the

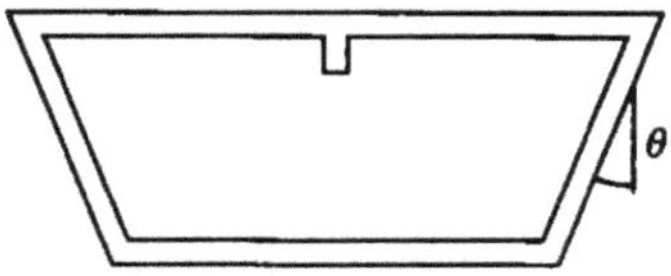

Figure 8.13 Beam with inclined shear webs.

individual webs. In finding m_2, the quantity S_f should be shared between the webs.

8.3.7 Inclined webs

For a beam with inclined webs, as in Figure 8.13, the various expressions for obtaining the shear force resistance of the section need to be adapted as follows:

1. *Yield check, method* 1. Equation (8.9) remains valid taking D as the overall section depth. t_1 is based on the actual metal thickness t, or $k_{z1}t$ if welded.
2. *Yield check, method* 2. Expressions (8.10) and (8.11) for V_c should be multiplied by cos θ, where θ is the angle of inclination of the webs, and t_2 is the actual metal thickness (or the effective thickness if welded).
3. *Buckling check.* Expression (8.12) or (8.16) for finding V_c should be multiplied by cos θ. In either expression, the depth d of the web plate is measured on the slope, and t is the actual metal thickness. Also d and t are defined in this way when obtaining p_{v1}, v_2, v_3, or m_1. It is necessary to multiply S_f by cos θ when calculating m_2.
4. *Web with tongue plates.* In applying equation (8.15), the expression defining V_{ct} should be multiplied by cos θ.

8.4 COMBINED MOMENT AND SHEAR

8.4.1 Low shear

At any given cross-section of a beam, it is usually found that moment and shear force act simultaneously. Obviously the designer must consider possible interaction between the two effects. In most beams, the moment is the critical factor, and the problem is to find whether or by how much the moment resistance is eroded by the presence of the shear.

If the applied shear force is reasonably small, the effect on the moment resistance is negligible and can be ignored. This is referred to as the 'low shear' case, and can be assumed to apply when the shear force V arising under factored loading does not exceed half the factored shear force resistance V_c/γ_m without moment.

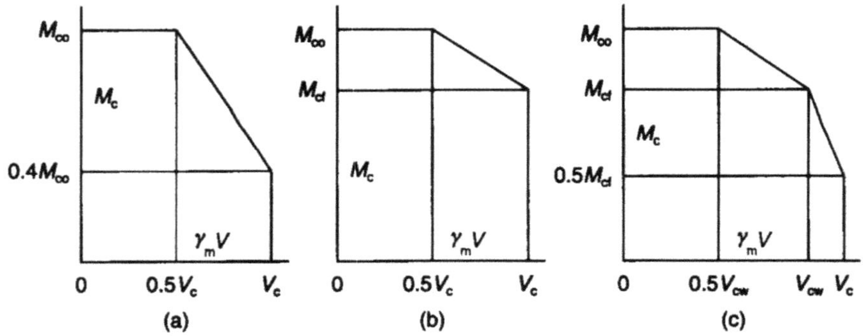

Figure 8.14 Moment/shear interaction: (a) method A; (b) method B; (c) with tension-field action.

8.4.2 High shear, method A

For the 'high shear' case ($V > 0.5V_c/\gamma_m$), the moment resistance must be suitably reduced to allow for the coexistent shear force. One method is to use a simple interaction diagram of the form shown in Figure 8.14(a), which gives the reduction in M_c as a function of V. In entering this, M_{co} is taken as the calculated moment resistance in the absence of shear. Such a diagram gives a reasonable result for solid bars (for which shear force is seldom a factor anyway), but is over-safe for typical beams containing thin webs.

8.4.3 High shear, method B

Figure 8.14(b) shows a more favourable form of interaction diagram, from which the reduced M_c may generally be found for members in which the shear is carried by webs. In entering this figure, M_{co} is again the value of M_c without shear, while M_{cf} is that part of M_{co} which is contributed by the flanges with web (and tongue plates) excluded.

When tension-field action has been included in the determination of V_c, Figure 8.14(b) becomes invalid and a diagram of the form shown in Figure 8.14(c) should be employed instead. In this, we take V_{cw} as the value of V_c that would be obtained by putting $m = 0$ in equation (8.16).

8.5 WEB CRUSHING

8.5.1 Webs with bearing stiffeners

At any position where a concentrated load or reaction acts on a beam, it is obviously desirable to provide a properly designed transverse stiffener, to prevent crushing of the web. The design of such stiffeners, known as 'bearing stiffeners', is covered in Section 8.6.3 below.

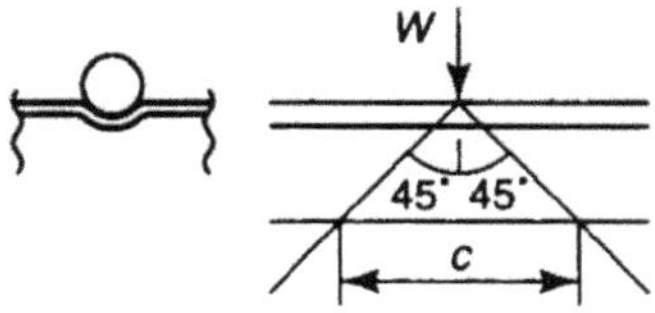

Figure 8.15 Web crushing.

8.5.2 Crushing of unstiffened webs

In the absence of a bearing stiffener, either for economy or because of a rolling load (Figure 8.15), it is necessary to consider the possibility of localized web crushing. The designer must determine the calculated resistance W_c to such failure, and ensure that this is not less than γ_m times the factored load W acting at the point considered, where W is the nominal (working) load or reaction at that point multiplied by γ_f. The BS.8118 procedure for finding W_c is to obtain two values, based on yield and on buckling, and take the lower. Both assume a 45° dispersion angle.

1. *Yield failure.* The critical position for this will typically be either adjacent to the flange or adjacent to a tongue-plate. For any given position, W_c may be calculated using the expression:

$$W_c = ct_w p_a \tag{8.17}$$

where: c = width of dispersion zone at the level considered,
t_w = actual metal thickness t at that level, if unwelded,
$= k_{z1}t$ in an HAZ region,
k_{z1} = HAZ softening factor (Section 6.4),
p_a = limiting stress for the material.

If necessary, W_c should be calculated at two or more levels, and the lowest value taken.

2. *Buckling failure.* For this, it is assumed that the load is carried by a vertical strip of the web acting as a strut, with cross-section $c_o t$ where c_o is the value of c at the neutral axis. W_c is found thus:

$$W_c = c_o t p_b \tag{8.18}$$

where t = actual web thickness, and p_b = limiting stress based on column buckling of the assumed strip. p_b may be read from a type C1 buckling curve defined by $p_1 = p_o$ (Section 9.5.2), taking the slenderness λ as follows:

$$\lambda = a(d/t) \tag{8.19}$$

where d is the web-plate depth between flanges or tongue plates if fitted. When the loaded flange is held from out-of plane movement relative to the other flange, a may be taken in the range 2.5–3.5, depending on the estimated degree of rotational restraint at the ends of the assumed strut. A value of $\alpha = 3.5$ would correspond to pinned ends. If however the loaded flange is free to deflect laterally as the web buckles, α should be suitably increased.

In a multi-web beam, W_c should be summed for all the webs. If the webs are inclined (Figure 8.13), the values of W_c given by equations (8.18) must be multiplied by $\cos\theta$, with the web-plate depth d measured on the slope.

8.6 WEB REINFORCEMENT

8.6.1 Types of reinforcement

The following types of reinforcement may be used for strengthening the webs of beams:

1. tongue plates;
2. bearing stiffeners (transverse);
3. intermediate stiffeners (transverse);
4. end-posts.

To be effective, these must be properly designed, the rules given below being based on BS.8118.

8.6.2 Tongue plates

The provision of tongue plates (Figure 8.16) can be beneficial in various ways. It enables web-to-flange welds to be moved in from the extreme fibres, thereby locating the HAZ in a region of lower stress. It reduces the d/t ratio of the web, thus improving the buckling performance in both bending and shear. Also, when a rolling load has to be carried on the top flange, it greatly increases the resistance to web crushing.

To be effective, a tongue should be proportioned so as not to be slender in terms of local buckling. A safe rule is to make the ratio of its depth b_t to its thickness t_t such that it would be classed as compact when considered

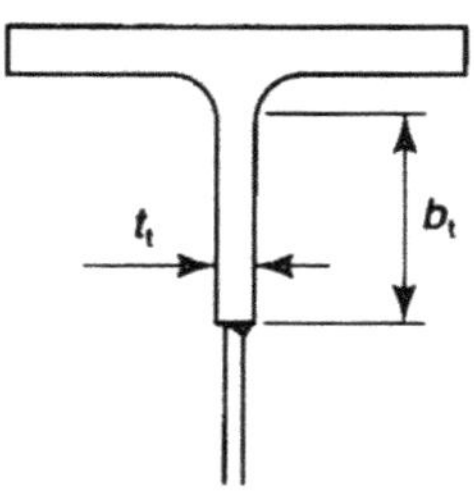

Figure 8.16 Tongue dimensions.

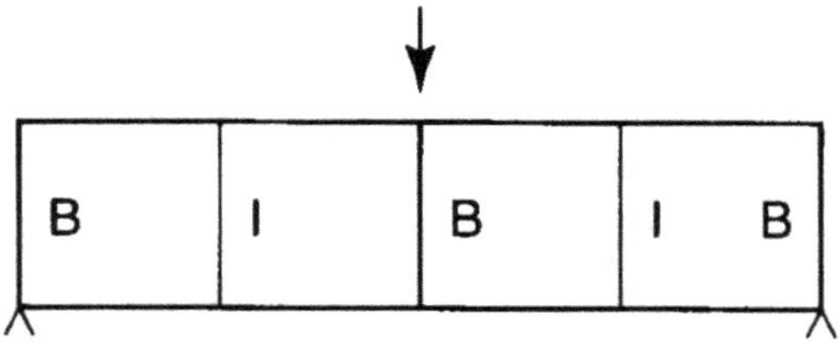

Figure 8.17 Transverse web stiffeners: bearing (B) and intermediate (I).

as an outstand element under uniform longitudinal compression:

$$\frac{b_t}{t_t} \nless 7\varepsilon \text{ (unwelded)} \qquad \text{or} \qquad 6\varepsilon \text{ (welded)} \tag{8.20}$$

8.6.3 Transverse stiffeners

Transverse web stiffeners may be one or other of two basic kinds:

- *Bearing stiffeners* are provided at points of concentrated load or reaction, to resist crushing;
- *Intermediate stiffeners* are fitted at intervals along the span, to divide the web into smaller panels and improve its resistance to shear buckling.

The stiffeners can be single or double-sided. They are often in the form of simple fins or outstands. Alternatively they can comprise channels applied in such a way as to create a closed section. When of the former type (a) the stiffener section ($b_s \times t_s$) should be such that it would be classed as compact when considered as an outstand element under compression (Section 7.1.4):

$$\frac{b_s}{t_s} \nless 7\varepsilon \text{ (unwelded)} \qquad \text{or} \qquad 6\varepsilon \text{ (welded)} \tag{8.21}$$

Transverse stiffeners must be designed to meet a strength requirement and also, generally, a stiffness requirement. In checking either of these, it is permissible to include an associated width of web-plate as part of the effective stiffener section, extending a distance b_1 each side of the actual stiffener (Figure 8.18), generally given by:

$$b_1 = \text{lesser of } 0.13a \text{ and } 15\varepsilon t \tag{8.22}$$

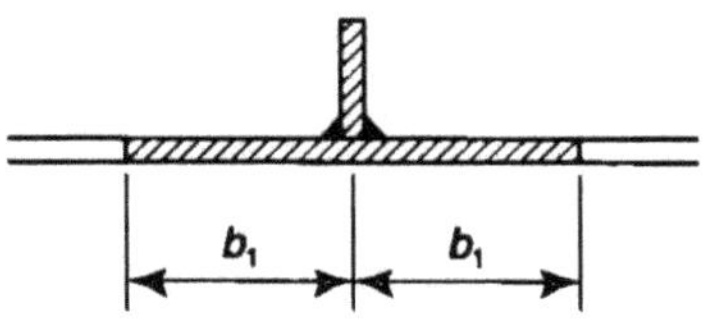

Figure 8.18 Transverse web stiffener, effective section (plan view).

where a is the stiffener spacing and t the web thickness. At the outboard side of an end stiffener, it should be taken as the lesser of a_0 and $6\varepsilon t$ where a_0 is the distance to the free edge of the web.

(a) Strength

The effective stiffener is checked as a strut, buckling out of the plane of the web, which is required to carry a notional load P under factored loading.

Bearing stiffener $$P = W + V/3 \tag{8.23a}$$

Intermediate stiffener $$P = V/3 \tag{8.23b}$$

where W = factored load or reaction applied to the beam at the stiffener considered, and V = average shear force arising in the panels either side of the stiffener, under factored loading.

The effective stiffener section, including the associated width of web plate, must be such that its calculated resistance P_c to column buckling about an axis parallel to the web is not less than $\gamma_m P$. P_c may be found as in Section 9.5, with suitable allowance for any HAZ softening within the effective section, and taking an effective strut length l generally as follows:

$$a/d > 1.5 \qquad l = d \tag{8.24}$$

$$a/d < 1.5 \qquad l = \frac{d}{\sqrt{(1.6 - 0.4a/d)}} \tag{8.25}$$

where d is the depth of the web plate between flanges or tongue plates, if fitted. When the panel dimension is different on the two sides of the stiffener, a should be taken as the mean value. If the applied load W acts eccentric to the axis of the stiffener, due account should be taken of the moment thereby introduced (combined axial load and moment).

(b) Stiffness

When a is less than $2.9d$, the following requirement should also be satisfied:

$$I_s \geq dt^3\left(\frac{2d}{a} - 0.7\right) \tag{8.26}$$

where I_s = inertia of the effective stiffener section, including the associated part of the web, about an axis through its centroid parallel to the web, with HAZ softening ignored.

8.6.4 End-posts

An 'end-post' is a special kind of transverse web stiffener that may be needed at the ends of a deep slender beam: (a) to provide twist resistance at that point, and/or (b) to anchor a tension field in the end panel of the web. These functions are in addition to the normal one of acting as a bearing stiffener to resist the reaction force. An end-post should be double-sided, with stiffening material added symmetrically either side of the web. It should extend the full depth between flanges.

(a) Twist resistance

This function of an end-post is important in relation to lateral-torsional buckling. A deep slender beam will be less prone to this if the top flange is positively held from moving sideways over the points of support. One way is to properly secure the bottom flange at these positions, and prevent distortion of the web (and hence top flange movement) by providing a properly designed end-post. For such an end-post to be effective, the inertia I_{ep} of its section about the mid-plane of the web should satisfy the following requirement:

$$I_{ep} > \frac{d^3 t_f R}{250 \sum W} \tag{8.27}$$

where d = depth of web between flanges or tongue plates, t_f = thickness of top-flange, R = reaction at the end of the span considered under factored loading, and $\sum W$ = total factored loading on the span.

(b) Tension-field anchor

It is only permissible to take advantage of tension-field action in an end panel of a web, if a suitable end-post is provided to react against the tension-field force. This may comprise a conventional bearing stiffener over the reaction point, together with a second stiffener, such that with the included piece of web they effectively form a short vertical beam (EP in Figure 8.12). This 'beam', spanning between the girder flanges, has to resist the horizontal component of the pull from the tension field. It must be designed to resist the action-effects arising from this pull, namely a moment M and a shear-force V both acting in the plane of the web. The values of these under factored loading may be estimated as follows (based on the Cardiff work):

$$V = 0.6dtp_v \sqrt{\left(1 - \frac{p_{v1}(q - p_{v1})}{v_2 p_v^2}\right)} \tag{8.28}$$

$$M = 0.1dV \tag{8.29}$$

where: q = mean shear stress acting in the end-panel of the web (under factored loading), based on the full area of the web-plate and tongue-plates (if fitted), p_v = limiting material stress for the web material in shear, p_{v1} = buckling stress without tension-field action (Figure 8.11), and v_2 = tension-field parameter (Section 8.3.6).

The section of the end-post must be adequate to resist M and V simultaneously, when checked as in Section 8.4.

8.7 LATERAL-TORSIONAL BUCKLING

8.7.1 General description

When the compression flange of a deep beam is inadequately stabilized against sideways movement, there is a danger of premature failure due to lateral-torsional (LT) buckling, before the full moment resistance of the cross-section has been reached (Figure 8.19). The compression flange deflects sideways, dragging the tension flange with it, thus producing a buckling mode that is a combination of lateral and torsional deformation [15, 27].

As a very rough guide, it may be assumed that LT buckling will not be critical if lateral supports to the compression flange are provided at a spacing less than $40\varepsilon r_y$, where r_y is the minor axis radius of gyration of the section, and $\varepsilon = \sqrt{(250/p_o)}$. Nor will it be a factor for hollow shapes, such as box sections, unless these are very deep and narrow. Generally, it is found that LT buckling becomes serious for deep non-hollow shapes, especially when the metal is thin.

If it is possible that LT buckling might be a critical factor, the adequacy of any proposed design may be checked using the procedure given below. The basic method considers the case of an applied moment acting about the major principal axis of the section. If there is also some moment about

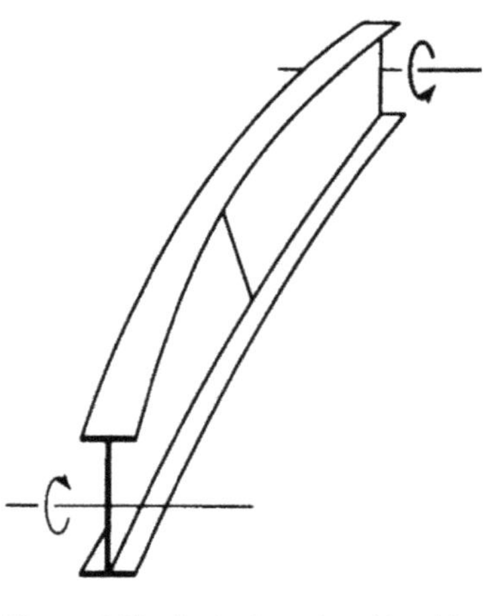

Figure 8.19 Lateral-torsional buckling.

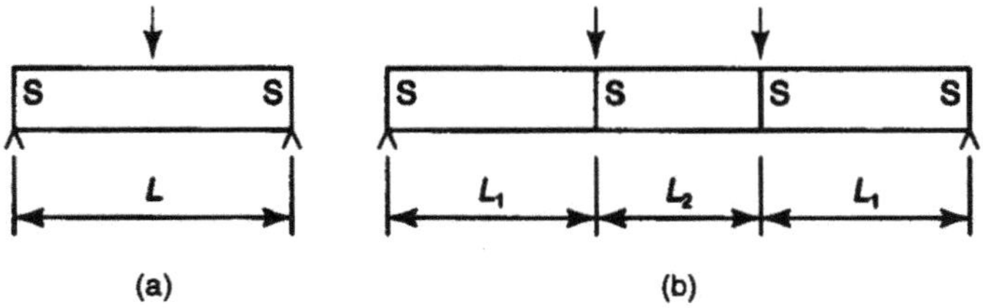

Figure 8.20 Lateral-torsional buckling: (a) buckling occurs in a single bay over the whole length (L) of the beam, no lateral support being provided by the load; (b) buckling occurs in one of the bays (L_1 or L_2) between points of lateral support. S = lateral support.

the minor axis ('side moment'), the combined effect of the two components should be allowed for by using the interaction equation given in Section 8.7.9

If effective lateral support is provided only at the end reaction points, with the compression flange otherwise free to deflect sideways, buckling is investigated by treating the whole span as a single unit or bay (Figure 8.20(a)). Alternatively if one or more lateral supports are provided along the length of the span, then the stability of each bay between such points of support must be checked (Figure 8.20(b)). In this case, the designer should also establish that these points really do provide lateral support to the compression flange, if necessary by employing a properly designed plan-view bracing system. The checking procedure we give is broadly based on BS.8118.

8.7.2 Basic check

In order to check against possible failure by LT buckling, it is necessary to ensure that the following requirement is met in every bay between points of effective lateral support:

$$\overline{M} < \frac{Sp_b}{\gamma_m} \tag{8.30}$$

where: $\overline{M}$ = equivalent uniform moment in the bay considered,
S = plastic modulus of *gross* section about major principal axis,
p_b = LT buckling stress (Section 8.7.4),
γ_m = material factor (Section 5.1.3).

8.7.3 Equivalent uniform moment

$\overline{M}$ is the magnitude of an assumed uniform moment that would have the same effect, in terms of buckling, as the actual pattern of moment on the bay considered. In a preliminary check, the designer can conservatively take $\overline{M}$ equal to the greatest moment arising in the bay. For a more accurate value, it is necessary to consider the shape of the bending moment diagram. Two cases arise, namely whether or not intermediate loading acts on the beam between points of lateral support.

(a) No intermediate loads

This case refers to a bay in which the bending moment diagram consists of a straight line (Figure 8.21(a)). It would typically apply when the loads on a beam are applied in such a way as to provide lateral support at the points where they act. For a bay under uniform moment, $\bar{M}$ is simply taken as the moment arising. If there is a moment gradient, $\bar{M}$ is taken as the greater of two values (A, B) found thus:

(A) $\bar{M}$ = moment arising at a point 40% of the way along the bay from the end with the higher moment (referred to as end 1);

(B) $\bar{M} = 0.4M_1$ where M_1 is the moment arising at end 1. This value will be found to govern when the moment at end 2 is such that it produces reverse curvature, and exceeds $0.5M_1$ in magnitude.

(b) Bay with intermediate loading

This case applies when loads act between positions of lateral support, in such a way that they themselves are free to move laterally as the beam buckles (Figure 8.21(b)). We suggest that for this case $\bar{M}$ be taken as the greater of two values (C, D) found thus:

(C) $\bar{M}$ = average moment arising in the central third of the bay, under factored loading;

(D) $\bar{M}$ = 60% of the greatest moment arising in the bay.

8.7.4 Limiting stress for LT buckling

The stress p_b in equation (8.30) may be read from the appropriate LT buck-

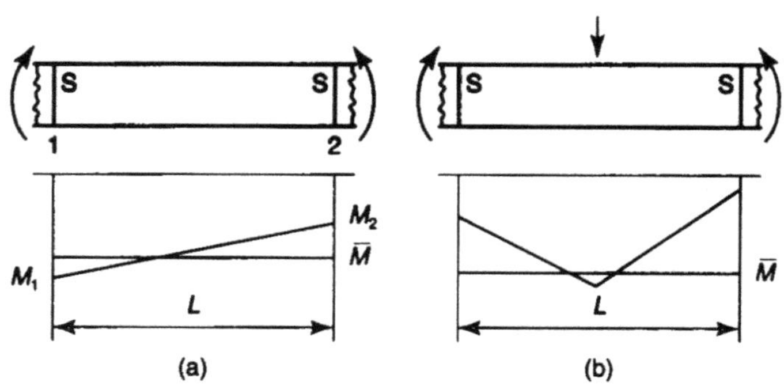

Figure 8.21 Lateral-torsional buckling in a bay between lateral supports (S), equivalent uniform moment $\bar{M}$: (a) bay with no intermediate loads; (b) bay with intermediate loading that fails to provide a lateral support.

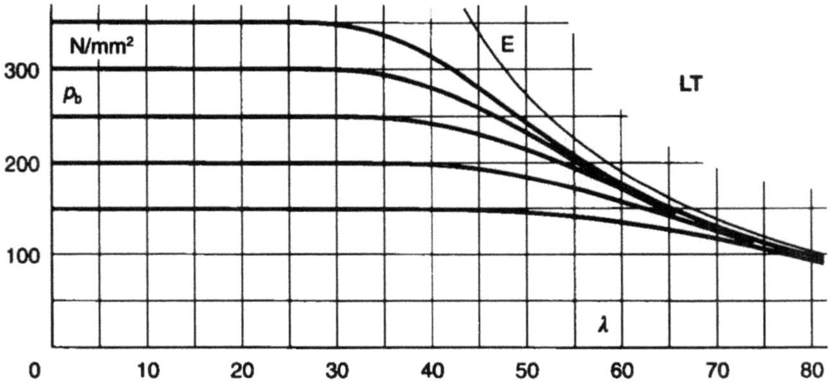

Figure 8.22 Limiting stress p_b for lateral-torsional buckling.

ling curve in Figure 8.22, defined by the value p_1 at which it meets the stress axis. For an ordinary extruded section, of fully compact section and unwelded, the relevant curve is selected by taking p_1 equal to the limiting stress p_o for the material. For any other section it is found thus:

$$p_1 = \frac{M_c}{S} \tag{8.31}$$

where M_c = calculated moment resistance of the cross-section (Section 8.2) making allowance for HAZ softening and local buckling, but ignoring small holes. If necessary, M_c should be suitably reduced to allow for the presence of shear (Section 8.4). S is the plastic modulus of the gross section.

For a welded beam, M_c should generally be based on the HAZ pattern at the least favourable cross-section within the bay considered. This applies even when the welds are localized, as with the attachment of transverse stiffeners, etc. However, when HAZ softening occurs only at the very ends of a bay, its presence may be ignored when considering LT buckling in that bay.

The LT curves in Figure 8.22 are based on equation (5.7) which may if desired be employed for the direct calculation of p_b.

8.7.5 Slenderness parameter

The determination of the slenderness parameter λ needed for entering the LT buckling curve tends to be laborious. For a preliminary check, it can very conservatively be taken equal to the effective slenderness λ_y ($= l/r_y$) for column buckling of the member about its minor axis. When a more realistic value is required, one of the methods given below can be used in which the required section properties may be found with the aid of Chapter 10.

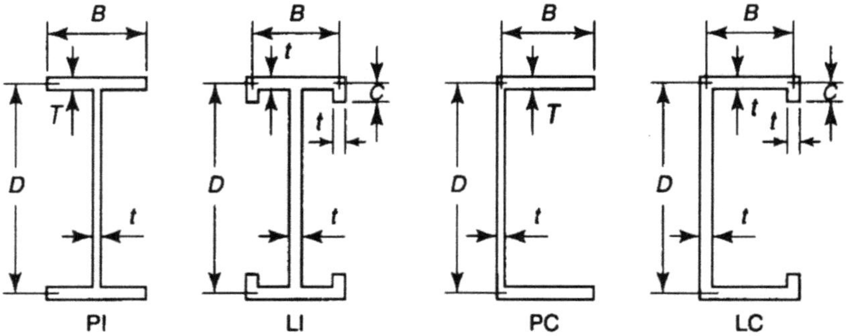

Figure 8.23 Sections covered in Table 8.1.

1. *Conventional I and channel sections*

For the sections covered by Figure 8.23 an acceptable estimate of λ can be found using the following formula:

$$\lambda = \frac{X\lambda_y}{\left\{1 + Y\left(\frac{\lambda_y T}{D}\right)^2\right\}^{1/4}} \tag{8.32}$$

where: D = depth of section,
T = flange thickness = t for sections LI and LC,
$\lambda_y = l/r_y$
l = effective buckling length of compression flange,
r_y = radius of gyration about minor axis for gross section.

The two coefficients may be safely taken as $X = 1.0$ and $Y = 0.05$. Alternatively, they can be calculated from the empirical formulae given in Table 8.1, leading to a better estimate for λ. Note that these formulae become unreliable when used outside the ranges indicated.

2. *Other sections*

For these λ should be found from the general expression:

$$\lambda = \pi\sqrt{\frac{ES}{M_{cr}}} \tag{8.33}$$

where M_{cr} = elastic critical moment, S = plastic modulus of the gross section, and E = modulus of elasticity = 70 kN/mm^2.

M_{cr} is the elastically calculated value of uniform moment at which LT buckling would occur in the bay considered, in the absence of any imperfections such as initial crookedness, HAZ softening, locked-in stress, etc. For symmetrical sections (Figure 8.24), M_{cr} may be calculated from

Table 8.1 Lateral-torsional buckling of certain conventional sections – empirical formulae for X and Y in equation (8.32)

Section	Formulae
PI. Plain I-section ($1 \leq \theta \leq 2$)	$X = 0.90 - 0.03\phi + 0.04\theta$ $Y = 0.05 - 0.010\sqrt{\phi(\theta - 1)}$
LI. Lipped I-section ($\theta = 1$)	$X = 0.94 - \phi(0.03 - 0.07\psi) - 0.3\psi$ $Y = 0.05 - 0.06\psi/\phi$
PC. Plain channel ($1 \leq \theta \leq 2$)	$X = 0.95 - 0.03\phi + 0.06\theta$ $Y = 0.07 - 0.014\sqrt{\phi(\theta - 1)}$
LC. Lipped channel ($\theta = 1$)	$X = 1.01 - \phi(0.03 - 0.06\psi) - 0.3\psi$ $Y = 0.07 - 0.10\psi/\phi$

Notes. 1. Refer to Figure 8.23 for section details.
2. Notation: $\phi = D/B$ $\psi = C/B$ $\theta = T/t$ for PI, PC $= 1.0$ for LI, LC.
3. The formulae are only valid when: $1.5 \leq \phi \leq 4.5$ $\psi \leq 0.5$ $1 \leq \theta \leq 2$

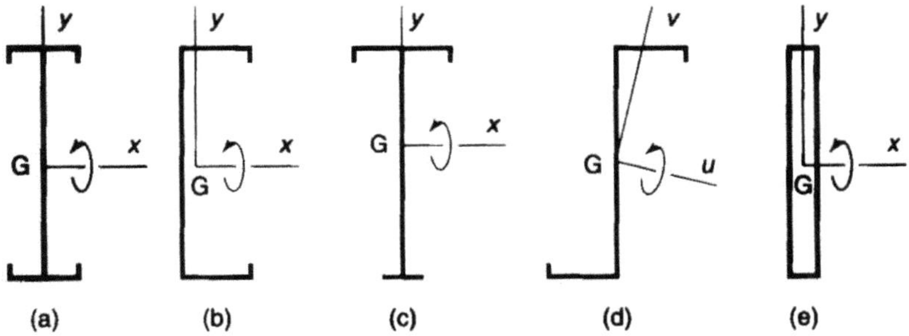

Figure 8.24 Lateral-torsional buckling, cases covered in Section 8.7.5 (2) and (3).

expressions given below, in which the various section properties should generally be based on the gross section.

(a) Bisymmetric section:

$$M_{cr} = \frac{1.9E}{l} \sqrt{\left\{ I_{yy}\left(\mathfrak{J} + \frac{26H}{l^2} \right) \right\}} \tag{8.34}$$

(b) Section symmetric about major axis: Same expression as for the bisymmetric section.

(c) Section symmetric about minor axis:

$$M_{cr} = \frac{1.9E}{l} \left[\sqrt{\left\{ I_{yy}\left(\mathfrak{J} + \frac{26H + 6\beta_x^2 I_{yy}}{l^2} \right) \right\}} + \frac{2.5\beta_x I_{yy}}{l} \right] \tag{8.35}$$

(d) Skew-symmetric section: Equation (8.34) may be used, but with I_{yy} changed to I_{vv}. The value obtained for M_{cr} relates to a moment acting about the major principal axis uu.

where: I_{yy}, I_{w} = inertia about minor principal axis,
$\mathfrak{J}$ = St Venant torsion factor (Section 10.4.2),
H = warping factor (Sections 10.4.5, 10.5),
β_x = special LT buckling factor (Section 10.4.6),
E = modulus of elasticity = 70 kN/mm^2.

3. *Hollow box-type sections*

These only need to be considered for LT buckling when they are especially deep in relation to their width (Figure 8.24(e)). Such sections have a very large torsion factor $\mathfrak{J}$, compared to an open section of similar size. It is therefore acceptable in their case to ignore warping effects and obtain M_{cr} from the following simplified expression:

$$M_{cr} = \frac{1.9E\sqrt{(I_{yy}\mathfrak{J})}}{l} \tag{8.36}$$

In order to obtain M_{cr} (and hence λ) for other section shapes the reader can refer to the book *Flexural-torsional Buckling of Structures* by Trahair [27].

8.7.6 Beams with very slender compression flanges

In considering the LT buckling of symmetrical I-beams and channels, in which the outstand element(s) forming the compression flange are very slender (Section 7.2.5), the question arises as to whether it is acceptable to exploit the post-buckled strength of such elements. There are two possible options:

1. *Method A* is effectively what BS.8118 does. Post-buckled strength is ignored in the determination of M_c, using expression (7.8) to obtain a_o for these elements. But in determining λ (Section 8.7.5) we base M_{cr} in equation (8.33) on the gross section. This method tends to become oversafe at low λ.
2. *Method B* is more favourable when the lateral supports are close spaced (low λ). Here M_c is based on an effective section in the compression flange that does take advantage of post-buckled strength, using expression (7.7) to obtain a_o. For convenience, the tension flange may be assumed to be reduced in the same way to preserve major axis symmetry. Here λ is determined as in Section 8.7.5(1), with I_{yy} based on the same effective section as that used for finding M_c, but with $\mathfrak{J}$ and H based on the gross section.

8.7.7 Effective length for LT buckling

In any given bay, the effective length l would normally be taken equal to the actual distance L between points of lateral support to the compression flange. This is valid when the flange, although held in position at these points, is free in terms of plan-view rotation. It is permissible to reduce the value of l when the flange is restrained against such rotation, although by just how much is a matter of some guesswork. British Standard BS.8118 permits $l = 0.85L$ for the case of 'partial' rotational restraint and $0.7L$ for 'full' restraint.

There are some situations where l should be taken greater than the bay length L, namely as follows:

(a) Destabilizing loads

For a bay that carries intermediate loading (Section 8.7.3(b)) it is important to consider at what height this loading acts. A destabilizing load is one which effectively acts at a point X located higher than the shear centre S (Figure 8.25). As the beam buckles, taking the load with it, the resulting moment about S aggravates the tendency to twist. Such a load therefore represents a more adverse condition than that of a non-destabilizing load (acting at, or below S), and must be allowed for by increasing l. Just how much depends on h (the height of X above S), a reasonable value being:

$$l = L\left(1 + \frac{0.4h}{D}\right) \tag{8.37}$$

where D is the overall depth of section. When the load is applied at top flange level in a symmetrical I-beam ($h = D/2$), this gives $l = 1.2L$. Equation (8.37) assumes that the compression flange is unrestrained against plan-view rotation at the ends of the bay. When restraint against such rotation is provided, BS.8118 permits effective lengths of L and $0.85L$ for the case $h = D/2$, corresponding to 'partial' and 'full' restraint respectively.

(b) Beams without end-posts

It is desirable to stabilize the top flange of a beam against out-of-plane

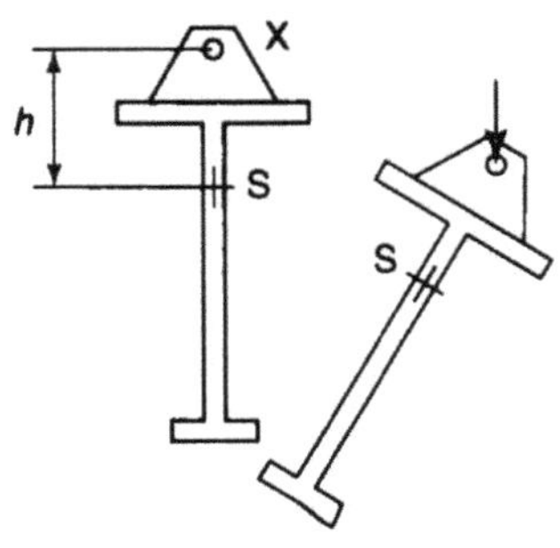

Figure 8.25 Lateral-torsional buckling, destabilizing load.

movement over the reaction points, by fitting a properly designed end-post (Section 8.6.4) or by some other means. Not so doing will reduce the resistance of an end bay to LT buckling, since some lateral movement of the compression flange over the point of support will then become possible (accompanied by distortion of the web). For a beam that is free to buckle over its whole length, the absence of end-posts can be allowed for by increasing l by an amount $2D$ above the value that would otherwise apply.

(c) Cantilevers

The LT buckling of cantilever beams is a complicated subject, and beyond the scope of this book. The designer is referred to BS.8118, where it will be seen that the effective length for such a member can be anywhere between $0.5L$ and $7.5L$, or he/she may refer to Trahair's book [27].

8.7.8 Beams of varying cross-section

The above treatment of LT buckling has assumed the member to be of uniform section over the length of the bay considered. We now consider the case where the section is non-uniform, either with a tapering depth or else with a varying flange area. A simple procedure for such beams is to employ the same basic equation (8.30), but with the relevant quantities evaluated as follows:

1. S is based on the cross-section at X, the point in the bay where the maximum moment acts.
2. In selecting the appropriate buckling curve, p_1 in equation (8.31) is based on the section at X.
3. In entering the buckling curve to obtain p_b, the slenderness λ is obtained according to Section 8.7.5 using the section properties at X, but with the effective length l found as in (4) below.
4. l is taken as the value found in Section 8.7.7 for a beam of uniform section multiplied by a factor K:

   (a) When the variation in section simply consists of a variation in depth, along the length of the beam, the flange areas remaining constant, it is accurate enough to take $K = 1$.

   (b) If, however, the flange areas vary along the length, K (> 1) should be found thus:

   $$K = 1.5 - 0.5R_f \tag{8.38}$$

   where R_f is the ratio of minimum to maximum flange area along the length, based on the average for the two flanges when these differ.

5. $\bar{M}$ in equation (8.30) is taken as the maximum moment arising in the bay (under factored loading).

8.7.9 Effect of simultaneous side moment

Any tendency of a beam to fail by LT buckling, due to major axis moment, will be aggravated if minor axis moment acts too. In such a case the member may be checked using the following requirement taken from BS.8118:

$$\frac{\overline{M}_x}{Sp_b} + \frac{\overline{M}_y}{M_{cy}} < \frac{1}{\gamma_m} \tag{8.39}$$

where: $\overline{M}_x, \overline{M}_y$ = equivalent uniform moments under factored loading,
M_{cy} = calculated minor axis moment resistance of the section,
S, p_b = as defined in Section 8.7.2.

Apart from this requirement, the cross-section must be able to resist the combined action of the moments M_x and M_y arising at any point on the span, treated as a case of unsymmetrical bending (Section 8.2.9).

As written, the above applies to bisymmetric and monosymmetric sections. It may also be used for a skew-symmetric shape, with x and y changed to u and v.

8.8 BEAM DEFLECTION

8.8.1 Basic calculation

So far this chapter has been concerned with strength (limit state of static strength). For aluminium beams, it is also important to consider stiffness (serviceability limit state) because of the relatively low elastic modulus. This consists of calculating the elastic deflection Δ_E under nominal loading and checking that it does not exceed the permitted value Δ_L (Section 5.1.4). For non-slender sections, Δ_E can be calculated from standard deflection formulae, taking $E = 70$ kN/mm^2 and the inertia I as that for the gross section. HAZ effects may be ignored.

In cases of unsymmetrical bending (Section 8.2.9), the designer must resolve the applied loading into components parallel to the two principal axes and calculate the corresponding components of deflection, which are then combined vectorially. Figure 8.26 illustrates the case of a typical Z-section under vertical loading, for which the horizontal deflection exceeds the vertical deflection.

8.8.2 Beam of slender section

When a beam has a cross-section classified as slender, it is possible that local buckling of elements on the compression side will reduce the effective stiffness and thus lead to increased deflection. The designer can allow

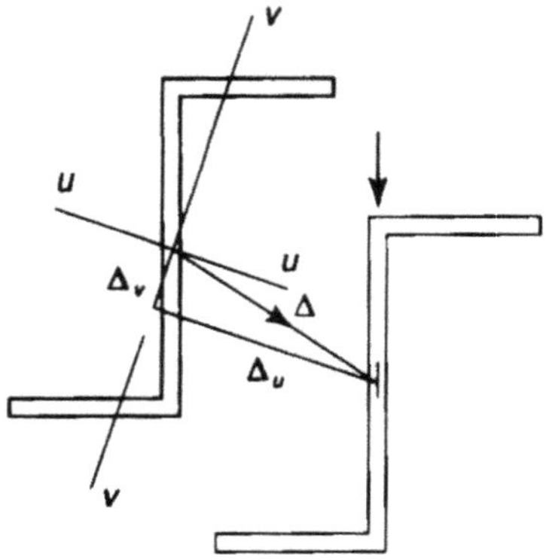

Figure 8.26 Components of deflection for asymmetric bending.

for this by taking a suitably reduced value of I, based on an effective section which allows for local buckling but ignores any HAZ effects.

For an initial check, the effective section can be found by determining the effective widths of any slender elements in the usual way (Chapter 7), putting $\varepsilon = \sqrt{(250/p_o)}$. However, such an approach tends to be pessimistic, leading to an overestimate of Δ_E, because it fails to allow for the lower level of stress under nominal as opposed to factored loading, which makes local buckling less critical. A more realistic result can be obtained by using a modified value for the parameter ε, equal to α times the usual value, where (equation (5.4)):

$$\alpha = \sqrt{(\gamma_m \gamma_f^*)} \qquad \text{or say } 1.25 \tag{8.40}$$

The section is re-classified using this value, and if it is now found to be non-slender Δ_E may be re-calculated using I for the gross section. If the section is still slender, a more favourable effective section can be taken generally in accordance with Chapter 7, but using the modified ε.

CHAPTER 9

Tension and compression members

9.1 GENERAL APPROACH

9.1.1 Modes of failure

This chapter covers the static design of members subjected to axial force either in tension ('ties') or in compression ('struts'). The basic requirement is that the factored resistance should not be less than the tensile or compressive force (action-effect) arising in the member under factored loading. The *factored resistance* is found by dividing the *calculated resistance* by the factor γ_m (Section 5.1.3). There are four possible modes of failure to consider in checking such members:

1. localized failure of the cross-section (Section 9.3);
2. general yielding along the length (Section 9.4);
3. overall column buckling (Section 9.5);
4. overall torsional buckling (Section 9.6).

Check 1, which applies to both ties and struts, must be satisfied at any cross-section in the member. It is likely to become critical when a particular cross-section is weakened by HAZ softening or holes. Checks 2, 3 and 4 relate to the overall performance of the entire member. Check 2 is made for tension members, and checks 3 and 4 for compression members. Check 4 is not needed for hollow box or tubular sections.

Most of the chapter is concerned with finding the calculated resistance P_c to each mode of failure, when the force on the member acts concentrically, i.e. through the centroid of the cross-section. We then go on to consider the case of members which have to carry simultaneous axial load and bending moment (Section 9.7), one example of this being when an axial load is applied eccentrically (not through the centroid).

9.1.2 Classification of the cross-section (compression members)

An early step in the checking of a compression member is to classify the section as compact or slender. If it is compact, local buckling is not a factor and can be ignored. If it is slender, local buckling will reduce the strength and must be allowed for.

The classification procedure is first to classify the individual plate elements comprising the section, by comparing their slenderness β with the limiting value β_s (Section 7.1.4). The classification for the section as a whole is then taken as that for the least favourable element. Thus for a section to be compact, all its elements must be compact. If one element is slender, then the overall cross-section is slender.

Refer to Chapter 7 for the definition of the plate slenderness β (Section 7.1.3 or 7.4.5), and also for the determination of β_s (Table 7.1).

9.2 EFFECTIVE SECTION

9.2.1 General idea

It is important to consider three possible effects which may cause local weakening in a member, namely HAZ softening at welds, buckling of thin plate elements in compression and the presence of holes. These are allowed for in design by replacing the actual cross-section with a reduced or effective one (of area A_e) which is then assumed to operate at full strength.

Reference should be made to Chapter 6 in dealing with HAZ softening, and Chapter 7 for local buckling. Chapter 10 gives advice on the determination of section properties.

9.2.2 Allowance for HAZ softening

Referring to Chapter 6, we assume that at any welded joint there is a uniformly weakened zone (HAZ) of nominal area A_z, beyond which a step-change occurs to full parent properties. We take an effective thickness of $k_z t$ in this zone and calculate A_e accordingly, with the softening factor k_z put equal to k_{z1} or k_{z2} depending on the type of resistance calculation being performed:

Local failure of the cross-section	$k_z = k_{z1}$
General yielding	$k_z = k_{z2}$
Overall buckling	$k_z = k_{z2}$

Alternatively a 'lost area' of $A_z(1 - k_z)$ can be assumed at each HAZ, which is then deducted from the section area. This procedure is often preferable at a cross-section just containing small longitudinal welds, as it avoids the need to find the actual disposition of the HAZ material, A_z for such welds being a simple function of the weld size (Section 6.5.6). The first method becomes necessary when there are transverse welds, and for bigger welds generally.

Care must be taken in dealing with a plate where the HAZ does not penetrate all the way through the thickness (Figure 6.16). The factor k_z need only be applied to the softened part of the thickness in such a case.

9.2.3 Allowance for local buckling

When a column cross-section has been classed as slender, the effective section of any slender element in it is assumed to be of the form shown in Figure 7.4 (internal elements) or Figure 7.5 (outstands), using an effective width model. The determination of the effective block-width (b_{e1} or b_{e0}) is explained in Sections 7.2.3 and 7.2.4, or in Section 7.4.7 for reinforced elements. Very slender outstands need special consideration (Section 9.5.4).

9.2.4 Allowance for holes

At any given cross-section, holes are generally allowed for by deducting an amount dt per hole, where d is the hole diameter and t the plate thickness. Exceptions are as follows:

1. For a hole in HAZ material, the deduction per hole need only be $k_z dt$.
2. Filled holes can be ignored in compression members.
3. A hole in the ineffective region of a slender plate element in a compression member can be ignored.

When a group of holes in a member is arranged in a staggered pattern, as for example in the end connection to a tension member, there are various possible paths along which tensile failure of the member might occur and it may not be obvious which is critical. Thus in Figure 9.1(a) the member might simply fail in mode A on a straight path through the first

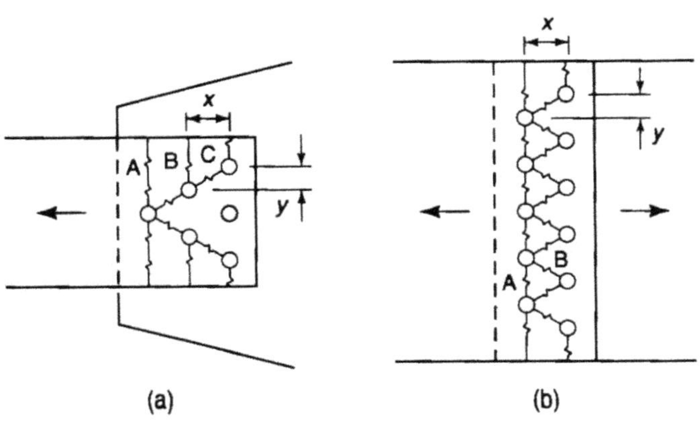

Figure 9.1 Staggered holes in tension members.

hole. Alternatively, it might fail in mode B or C, involving a crooked failure path with diagonal portions in it. Similarly in Figure 9.1(b). In such cases, the effective area A_e must be found for each possible failure path, and the lowest value taken. In considering a crooked failure path, A_e can be estimated as follows:

$$A_e = A - ndt + \sum \frac{x^2 t}{4y} \tag{9.1}$$

where A = total transverse section area, n = number of holes on the failure path, and x, y = longitudinal and transverse hole pitch, as shown. The summation is made for every diagonal portion of the failure path between pairs of holes.

9.3 LOCALIZED FAILURE OF THE CROSS-SECTION

The calculated resistance P_c of an axially loaded member to localized failure at a specific cross-section is found thus:

$$P_c = A_e p_a \tag{9.2}$$

where p_a = limiting stress for the material (Section 5.3), and A_e = effective section area.

The use of the higher limiting stress p_a, rather than the usual value p_o, reflects the view that some limited yielding at a localized cross-section need not be regarded as representing failure of the member.

The effective area A_e should be based on the most unfavourable position along the member, making suitable allowance for HAZ softening, local buckling (compression only) or holes as necessary (Section 9.2). When considering the local buckling of a very slender outstand, it is permissible to take advantage of post-buckled strength for this check (Section 7.2.5).

In the case of hybrid members containing different strength materials, P_c should be found by summing the resistances of the various parts, taking an appropriate p_a for each.

9.4 GENERAL YIELDING ALONG THE LENGTH

This is the form of failure in which the member yields all along its length, or along a substantial part thereof. It need only be checked specifically for tension members, because the column buckling check automatically covers it in the case of struts. The calculated resistance P_c is obtained thus:

$$P_c = A_e p_o \tag{9.3}$$

where p_o = limiting stress for the material (Section 5.3), and A_e = effective section area.

One difference from the treatment of localized failure is the use of the limiting stress p_o, rather than the higher value p_a. For most materials, p_o is taken equal to the 0.2% proof stress.

The other difference is that the effective area A_e now relates to the general cross-section of the member along its length, ignoring any localized weakening at the end connections or where attachments are made. For a simple extruded member, therefore, A_e may be taken equal to the gross area A. Holes need only be allowed for if there is a considerable number of these along the member. Likewise, a deduction for HAZ softening is only necessary when the member contains welding on a significant proportion of its length.

9.5 COLUMN BUCKLING

9.5.1 Basic calculation

Column (or flexural) buckling of a strut is the well-known mode of failure in which the central part of the member 'pops out sideways'. The calculated resistance P_c is given by:

$$P_c = Ap_b \tag{9.4}$$

where p_b = column buckling stress, and A = gross section area.

This equation should be employed to check for possible buckling about each principal axis of the section in turn. For a built-up member, consisting of two or more components connected together at intervals along the length, buckling should be checked not only for the section as a whole (about either axis), but also for the individual components between points of interconnection. For back-to-back angles, such that the buckling length is the same about both principal axes, accepted practice is to interconnect at third-points.

9.5.2 Column buckling stress

The buckling stress p_b depends on the overall slenderness λ. It may be read from one of the families of curves (C1, C2, C3) given in Figure 9.2, the derivation of which was explained in Chapter 5. Alternatively, it may be calculated from the relevant formula (Section 5.4.2).

The appropriate family, which need not necessarily be the same for both axes of buckling, is chosen as follows:

	Unwelded	*Welded*
Symmetric or mildly asymmetric section	C1	C2
Severely asymmetric section	C2	C3

The terms 'symmetric' and 'asymmetric' refer to symmetry about the axis

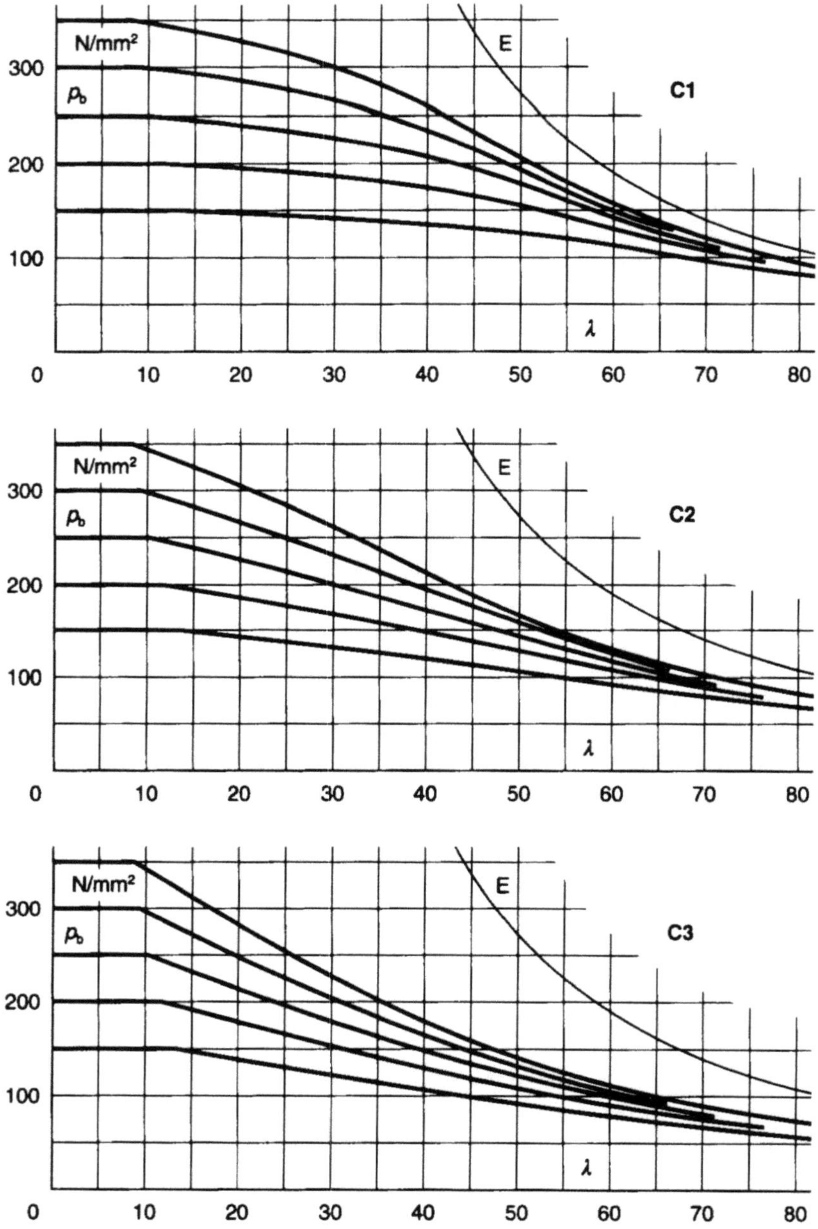

Figure 9.2 Limiting stress p_b for column buckling.

of buckling. A severely asymmetric section is one for which y_1 exceeds $1.5y_2$, where y_1 and y_2 are the distances from this axis to the further and nearer extreme fibres. For family selection, a strut is regarded as 'welded' if it contains welding on a total length greater than the largest dimension of the section.

The different curves in each family are defined by the stress p_1 at which they meet the stress axis. Having selected the right family, the appropriate curve in that family is found by taking p_1 as follows:

$$p_1 = \left(\frac{A_e}{A}\right) p_o \tag{9.5}$$

where p_o = limiting stress for the material (Section 5.3), A = gross section area, and A_e = effective section area (Section 9.2).

A_e relates to the basic cross-section, with weakening at end connections ignored. For a compact extruded member, $A_e = A$ and $p_1 = p_o$.

In finding p_1 for a welded strut, HAZ effects must generally be allowed for, even when the welding occupies only a small part of the total length. They can only be ignored when confined to the very ends. It is seen that welded struts are doubly penalized: firstly, in the use of a less favourable family and, secondly in the adoption of an inferior curve in that family (lower p_1). No deduction for unfilled holes need be made in the overall buckling check, unless they occur frequently along the length.

9.5.3 Column buckling slenderness

The slenderness parameter λ needed for entering the column buckling curve (C1, C2 or C3) is given by:

$$\lambda = \frac{l}{r} \tag{9.6}$$

where l = effective buckling length, and r = radius of gyration about the relevant principal axis, generally based on the gross section.

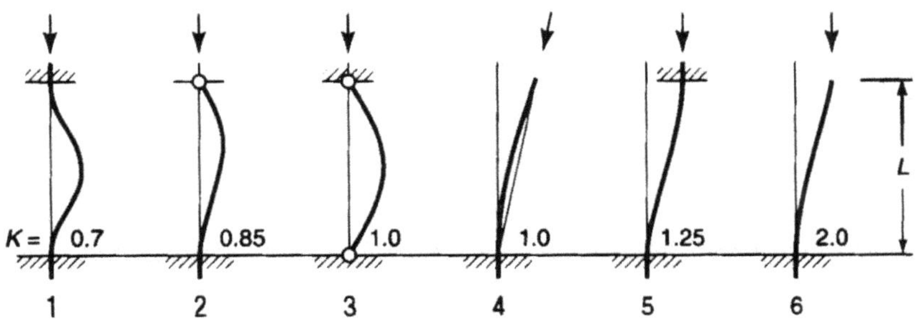

Figure 9.3 Column buckling, effective length factor K.

The determination of l involves a considerable degree of judgment (i.e. guesswork) by the designer – as in steel. It is found as follows:

$$l = KL \tag{9.7}$$

where L = unsupported buckling length, appropriate to the direction of buckling. K may be estimated with the help of Figure 9.3.

9.5.4 Column buckling of struts containing very slender outstands

For a column containing very slender outstands (Section 7.2.5), the designer must know whether it is permissible to assume an effective section that takes account of the post-buckled strength of these. There are two possible procedures:

1. *Method A* is effectively the same as that given in BS.8118. p_1 (expression (9.5)) is based on an effective section that ignores post-buckled strength in such elements, using expression (7.8) to obtain α_o. But in finding λ, it bases r on the gross section. It is further assumed that the member is under pure compression (no bending) when the applied load aligns with the centroid of the *gross* section.
2. *Method B* employs an effective section which takes advantage of post-buckled strength in very slender elements, with α_o found from expression (7.7). This effective section is employed for obtaining both A_e and r. The member now counts as being in pure compression when the applied load acts through the centroid of the *effective* section.

Method B thus employs a higher buckling curve (higher p_1), but enters it at a higher λ. Method A is found to be the more favourable for the majority of cases, but method B becomes advantageous if the member is short (low λ).

9.6 TORSIONAL BUCKLING

9.6.1 General description

There are three fundamental modes of overall buckling for an axially loaded strut [27]:

1. column (i.e. flexural) buckling about the minor principal axis;
2. column buckling about the major axis;
3. pure torsional buckling about the shear centre S.

Figure 9.4 shows the mid-length deflection corresponding to each of these for a typical member. The mode with the lowest failure load is the one that the member would choose.

Torsion needs to be considered for open (non-hollow) sections. Because the torsional stiffness of these is roughly proportional to thickness cubed,

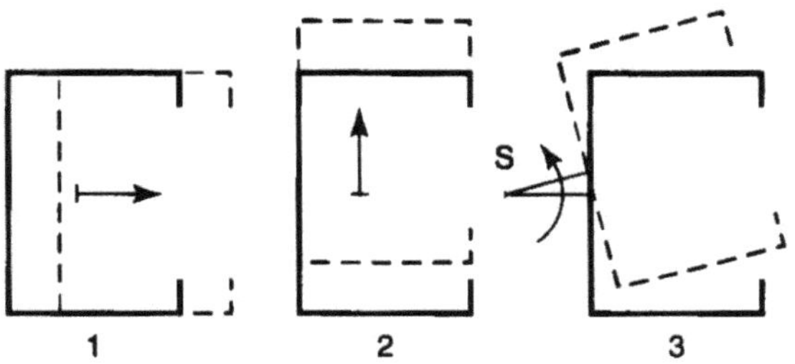

Figure 9.4 Fundamental buckling modes for a compression member.

the torsional mode never governs when the section is thick, as in heavy gauge (hot-rolled) steel. It becomes more likely as the thickness decreases, and for thin-walled members it is often critical, as also in light gauge (cold-formed) steel. The checking of torsional buckling tends to be laborious. Here we follow the treatment given in BS.8118, which is more comprehensive than any provided in previous codes. The calculations involve the use of torsional section properties, which may not be familiar to some designers. Chapter 10 provides help for evaluating these.

9.6.2 Interaction with flexure

A tiresome complication with torsional buckling is that the fundamental buckling modes often interact, depending on the degree of symmetry in the section (Figure 9.5):

1. *Bisymmetric section*. The three modes are independent (no interaction);
2. *Radial-symmetric section*. As for 1;
3. *Skew-symmetric section*. As for 1;
4. *Monosymmetric section*. Interaction occurs between pure torsional buckling about the shear centre S and column buckling about the axis of symmetry *ss*;
5. *Asymmetric section*. All three modes interact.

The effect of interaction is to make the section rotate about a point other than the shear centre S, as illustrated for a monosymmetric section in

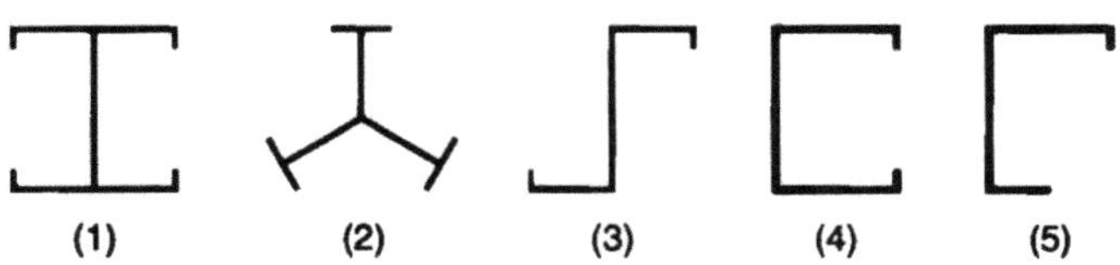

Figure 9.5 Degrees of symmetry for strut sections: (1) bisymmetric; (2) radial-symmetric; (3) skew-symmetric; (4) monosymmetric; (5) asymmetric.

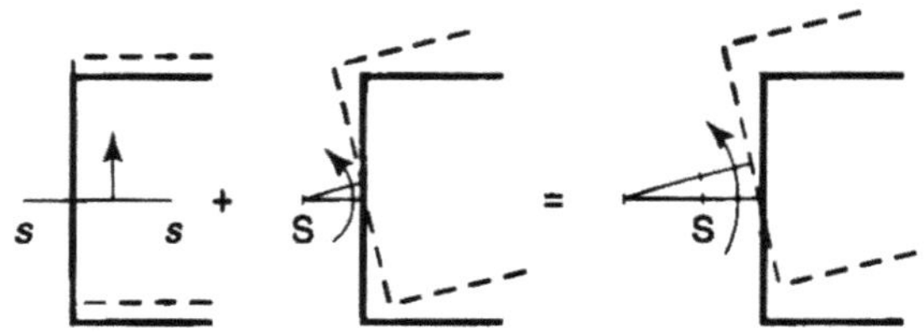

Figure 9.6 Monosymmetric section. Interaction between pure flexural buckling about *ss* and pure torsional buckling about S.

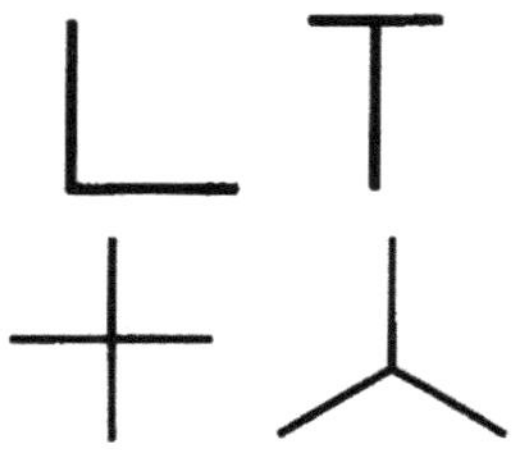

Figure 9.7 Typical 'type-R' sections, composed of radiating outstands.

Figure 9.6, and leading to a reduced failure load. Strictly speaking, interaction between torsional and column buckling can occur very slightly even with thick members. But the effect only becomes significant when the section is thin.

9.6.3 'Type-R' sections

In dealing with torsional buckling it is important to distinguish between 'type-R' sections and all others. A type-R section is one that consists entirely of radiating outstands, such as angles, tees and cruciforms (Figure 9.7). For such members, each component element is simply supported along the common junction, or nearly so. When such an element suffers local buckling, it typically does so in one sweep occupying the whole length of the member (Figure 9.8), and not in a localized buckle as for other thin-walled shapes. This is essentially a torsional mode of deformation. In thin type-R sections, therefore, local buckling and torsional buckling amount to much the same thing.

In design, it is convenient to treat the buckling of type-R struts in terms of torsion, rather than local buckling. By so doing, one is able to take advantage of the rotational restraint that the outstands may receive from the fillet material at the root. Double-angle (back-to-back) struts can also be regarded as effectively type-R.

9.6.4 Sections exempt from torsional buckling

Torsional buckling is never critical for a strut with any of the sections

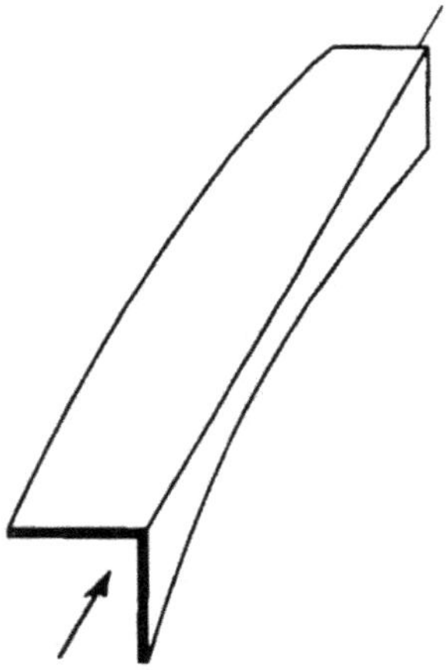

Figure 9.8 Torsional buckling of a type-R section strut.

listed below, and need not be checked:

1. Box or tubular sections (always much stiffer in torsion than a comparable open section);
2. Any type-R section that *would* be classified as compact for local buckling;
3. Conventional bisymmetric I-sections (with unreinforced flanges);
4. Conventional skew-symmetric Z-sections (with unreinforced flanges).

9.6.5 Basic calculation

As with column buckling, the basic expression for the calculated resistance P_c is as follows:

$$P_c = Ap_b \tag{9.8}$$

where p_b is now the torsional buckling stress, allowing for interaction with flexure if necessary, and A is the gross section area as before.

9.6.6 Torsional buckling stress

The buckling stress p_b depends on the torsional buckling slenderness parameter λ and may be read from one of the families of curves (T1, T2) given in Figure 9.9. Alternatively, it may be calculated from the relevant formula (Section 5.4.2). The appropriate family is chosen as follows, the T2 curves being the more favourable:

Type-R sections	T2
All other sections	T1

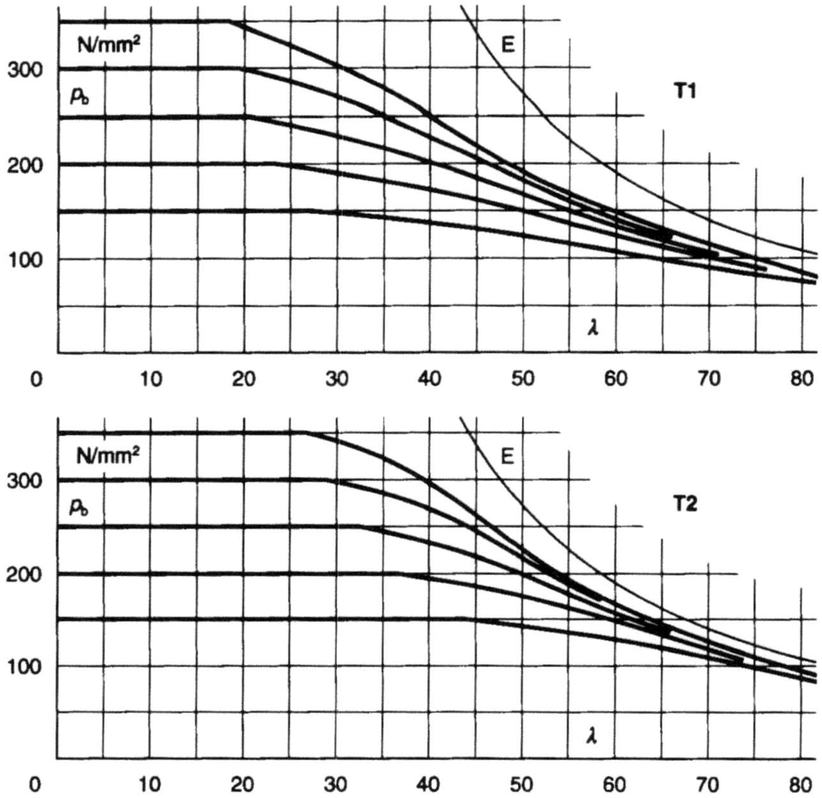

Figure 9.9 Limiting stress p_b for torsional buckling.

The different curves in each family are defined by the stress p_1 at which they cut the stress axis, the relevant curve being found by taking p_1 as follows:

$$p_1 = \left(\frac{A_e}{A}\right) p_o \tag{9.9}$$

where p_o = limiting stress for the material (Section 5.3), A = gross section area, and A_e = effective area of the cross-section, disregarding any localized weakening at the end-connections.

1. *Type-R sections.* A_e is found by making a suitable reduction to allow for any HAZ softening. *No reduction is made for local buckling.* Thus for an unwelded Type-R section $A_e = A$ even if its section is not compact.
2. *Other sections.* For these A_e is found in the same way as for column buckling, making suitable reductions to allow for both HAZ softening and local buckling. For a simple extruded member of compact section, $A_e = A$.

9.6.7 Torsional buckling slenderness

The following is the rigorous procedure for obtaining the slenderness parameter λ needed for entering the selected buckling curve. An alternative and quicker method, available for certain common shapes, involves the use of empirical formulae (Section 9.6.10). Under the rigorous procedure we take:

$$\lambda = k\lambda_t \tag{9.10}$$

where λ_t = slenderness parameter based on pure torsional buckling about the shear centre S, and k = torsion/flexure interaction factor (Section 9.6.8).

The slenderness parameter λ_t may be determined using the following general expression, which is valid for aluminium:

$$\lambda_t = \sqrt{\left(\frac{I_p}{0.0381\mathfrak{J} + H/l^2}\right)} \tag{9.11}$$

where $\mathfrak{J}$ = St Venant torsion factor, I_p = polar inertia about shear centre S, H = warping factor, and l = effective buckling length.

Here $\mathfrak{J}$, I_p and H may be based on the gross section, and can be found with the aid of Chapter 10. The effective length l is less critical than with column buckling, and is normally taken equal to the actual buckling length L. A lower value may be justified if there is significant warping restraint at the ends, but not if the ends are welded.

The warping factor H for type-R sections is always zero, or virtually so. It is therefore seen from the previous equation that λ_t for these is independent of length and becomes:

$$\lambda_t = 5.12\sqrt{\frac{I_p}{\mathfrak{J}}} \tag{9.12}$$

The torsional stability of type-R sections can be much improved by providing liberal bulb and/or fillet material, since this increases $\mathfrak{J}$. Fillet size is less important for non-R shapes, because these have warping resistance to improve their stability.

9.6.8 Interaction factor

The interaction factor k depends on the symmetry of the section (Figure 9.5) and should be found as follows, using the gross section in every case:

1. *Bisymmetric sections.* $k = 1$
2. *Radial-symmetric sections.* $k = 1$
3. *Skew-symmetric sections.* $k = 1$
4. *Monosymmetric sections.* k can be read from Figure 9.10 or else calculated from the corresponding formula:

$$k = \sqrt{\left(\frac{2Xs^2}{1 + s^2 - \sqrt{\{(1+s^2)^2 - 4Xs^2\}}}\right)} \tag{9.13}$$

where $s = \lambda_s/\lambda_t$

λ_s = slenderness parameter for column buckling about the axis of symmetry ss,

$X = (I_{ss} + I_{yy})/I_p$

I_{ss}, I_{yy} = inertias about the axis of symmetry and about the other principal axis, respectively.

5. *Asymmetric sections.* When the section has no symmetry, so that all three fundamental modes interact, k should be calculated as follows:

$$k = \sqrt{\frac{T}{x}} \tag{9.14}$$

where

$$T = \frac{3X}{1 + C/s_v^2} \qquad C = \left(1 - \frac{U^2}{r_p^2}\right) + \gamma\left(1 - \frac{V^2}{r_p^2}\right)$$

$$s_v = \frac{\lambda_v}{\lambda_t} \qquad X = \frac{I_{uu} + I_{vv}}{I_p}$$

$$r_p = \sqrt{\frac{I_p}{A}} \qquad \gamma = \frac{I_{uu}}{I_{vv}}$$

U, V = coordinates of shear-centre S (Figure 10.20),

I_{uu}, I_{vv} = major and minor principal axis inertias,

λ_v = minor axis column buckling slenderness,

λ_t = pure torsional buckling slenderness (equation (9.11)).

The quantity x is the lowest root of the following cubic, for the solution of which BS.8118 provides a convenient nomogram:

$$x^3 - 3x^2 + Ax - B = 0 \tag{9.15}$$

where

$$A = \frac{9X\{\gamma + s_v^2(1 + \gamma)\}}{(C + s_v^2)^2} \qquad B = \frac{27\gamma X^2 s_v^2}{(C + s_v^2)^3}$$

9.6.9 Torsional buckling of struts containing very slender outstands

When the section contains very slender outstands (Section 7.2.5), the question arises as to how to handle the local buckling thereof. Which effective section should be assumed: one that exploits the post-buckled strength of such elements, or one that ignores it? As with column buckling there are two possible options:

1. *Method A* is that adopted in BS.8118. p_1 is based on an effective section

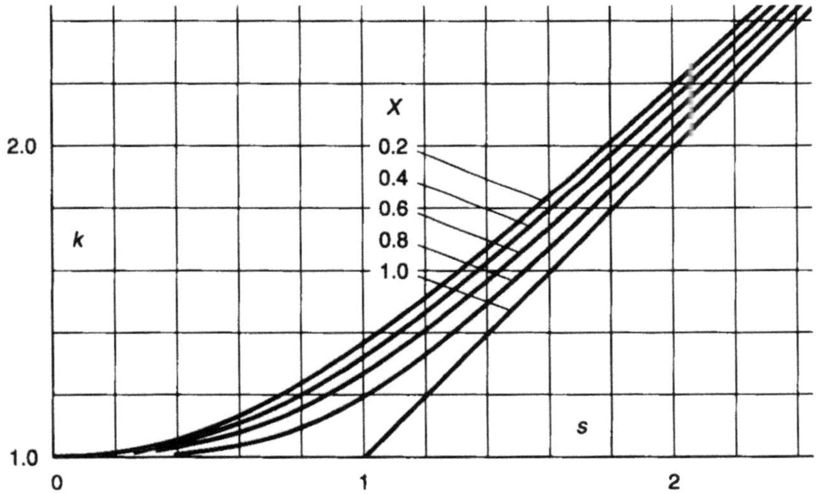

Figure 9.10 Torsional buckling of monosymmetric sections, interaction factor k.

that ignores the post-buckled strength of very slender outstands. But in finding λ, it bases λ_t and k on the gross section.

2. *Method B* employs an effective section which takes advantage of the post-buckled strength of these elements, this effective section being used for the determination of p_1. In finding λ, the method bases λ_s or λ_v on the same effective section, but all the other quantities are based on the gross section.

As with column buckling, method B tends to be more favourable than method A at low λ.

9.6.10 Empirical slenderness formulae

The torsional buckling slenderness parameter λ may be obtained more simply for some conventional sections by using empirical formulae. These are taken from BS.8118 and relate to the shapes shown in Figures 9.11 and 9.13. The slenderness λ is still found from the basic expression (9.10), but with λ_t now calculated using the appropriate data in Table 9.1 or 9.2. As before, k is read from Figure 9.10. Note that the formulae in the two tables will be inaccurate if used outside the ranges indicated.

1. *Type-R sections*. Figure 9.11 shows the shapes covered and the notation, the required formulae for λ_t being given in Table 9.1. The parameter ρ in these is a measure of the fillet reinforcement defined as follows (Figure 9.12):

$$\text{Radius-fillet: } \rho = \frac{R}{t} \qquad \text{Bevel-fillet (45°): } \rho = \frac{1.7F}{t}$$

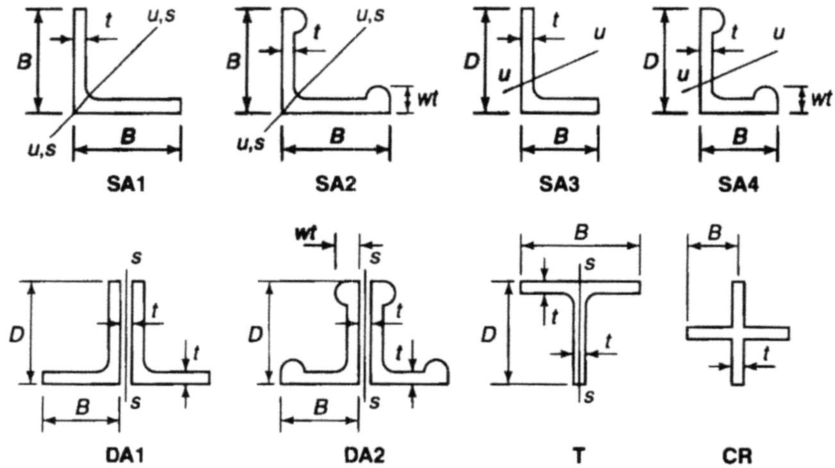

Figure 9.11 Type-R sections covered in Table 9.1.

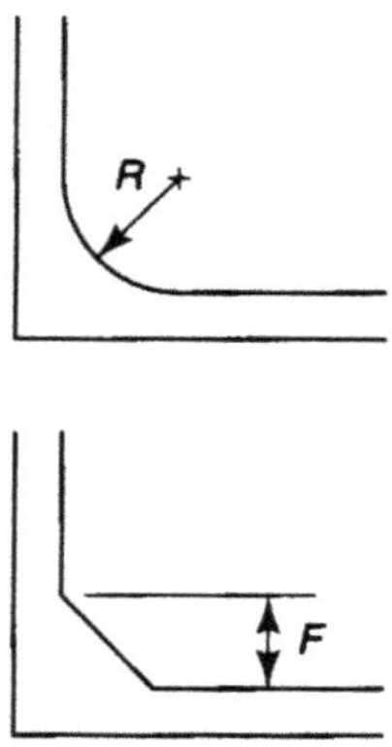

Figure 9.12 Root reinforcement.

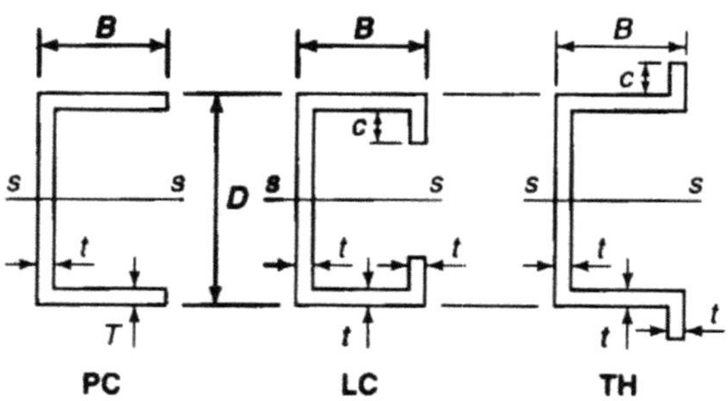

Figure 9.13 Sections covered in Table 9.2.

Table 9.1 Torsional buckling of certain type-R sections – empirical formulae for λ_t and X

Profile	Empirical formulae	Valid range
SA1	$\lambda_t = 5B/t - 0.6\sqrt{\rho^3 B/t}$ $X = 0.6$	$\rho < 5$
SA2	$\lambda_t = 5B/t - 0.6\sqrt{\rho^3 B/t} + 1.5\rho(w-1) - 2(w-1)^3$ $X = 0.6$	$\rho < 5$ $1 < w < 2.5$
SA3[2]	$\lambda_t = (D/t)\{4.2 + 0.8(B/D)^2\} - 0.6\sqrt{\rho^3 D/t}$ $X = 0.6 - 0.4(1 - B/D)^2$	$\rho < 5$ $0.5 < B/D < 1$
SA4[2]	$\lambda_t = (D/t)\{4.2 + 0.8(B/D)^2\} - 0.6\sqrt{\rho^3 D/t}$ $+ 1.5\rho(w-1) - 2(w-1)^3$ $X = 0.6 - 0.4(1 - B/D)^2$	$\rho < 5$ $0.5 < B/D < 1$ $1 < w < 2.5$
DA1	$\lambda_t = (B/t)\{4.4 + 1.1(D/B)^2\} - 0.7\sqrt{\rho^3 B/t}$ $X = 1.1 - 0.3D/B$	$\rho < 5$ $0.5 < D/B < 2$
DA2	$\lambda_t = (B/t)\{4.4 + 1.1(D/B)^2\} - 0.7\sqrt{\rho^3 B/t}$ $+ 1.5\rho(w-1) - 2(w-1)^3$ $X = 1.1 - 0.3D/B$	$\rho < 5$ $0.5 < D/B < 2$ $1 < w < 2.5$
T	$\lambda_t = (D/t)(1.4 + 1.5B/D + 1.1D/B) - \sqrt{\rho^3 D/t}$ $X = 1.3 - 0.8D/B + 0.2(D/B)^2$	$\rho < 3.5$ $0.5 < D/B < 2.5$
CR	$\lambda_t = 5.1B/t - \sqrt{\rho^3 B/t}$ $X = 1$	$\rho < 3.5$

Notes. 1. Refer to Figure 9.11 for section details.
2. Despite the asymmetry of sections SA3 and SA4 the interaction factor k is still found accurately enough using Figure 9.10. In entering this we take $s = (\lambda_u/\lambda_t)\{1 + 6(1 - B/D)^2\}$, where λ_u is the slenderness for column buckling about the major principal axis.

Table 9.2 Torsional buckling of certain channel-type sections – empirical formulae for ϕ, θ, X

Profile	Empirical formulae	Valid range
PC	$\phi = (B/T)\{7 + 1.5(D/B)(T/t)\}$ $\theta = 0.14 - 0.02D/B - 0.02T/t$ $X = 0.38D/B - 0.04(D/B)^2$	$1 < D/B < 3$ $1 < T/t < 2$
LC	$\phi = (B/t)\{7 + 1.5(D/B) + 5c/B\}$ $\theta = 0.12 - 0.02D/B + \dfrac{0.6(c/B)^2}{D/B - 0.5}$ $X = 0.38D/B - 0.04(D/B)^2 - 0.25c/B$	$1 < D/B < 3$ (uniform thickness)
TH	$\phi = (B/t)\{7 + 1.5(D/B) + 5c/B\}$ $\theta = 0.12 - 0.02D/B - \dfrac{0.05(c/B)}{D/B - 0.5}$ $X = 0.38D/B - 0.04(D/B)^2$	$1 < D/B < 3$ (uniform thickness)

Note. Refer to Figure 9.13 for section details.

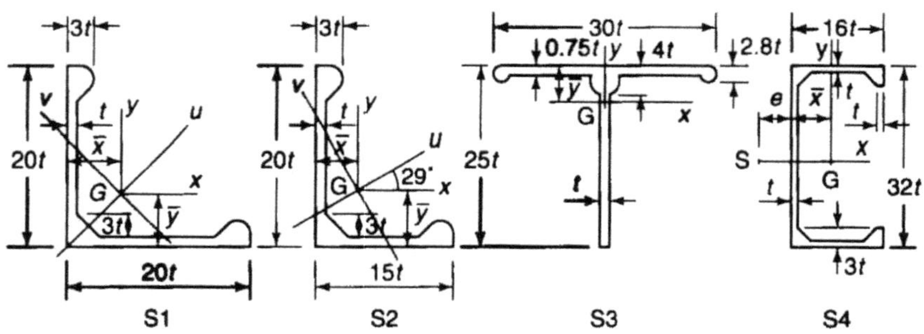

Figure 9.14 Four standardized profiles (BS. 1161).

The value of X, needed to obtain k, may be calculated using the relevant formula in Table 9.1. s is taken as in Section 9.6.8(4), except in the case of the unequal angles SA3 and SA4.

2. *Channel and top-hat sections.* The shapes covered and the notation, are shown in Figure 9.13. For all of these λ_t is calculated from the general expression:

$$\lambda_t = \frac{\phi}{\sqrt{(1 + \theta\phi^2/\lambda_s^2)}} \tag{9.16}$$

where λ_s = slenderness parameter for column buckling about ss, and

Table 9.3 Properties of certain standardized shapes

	Equal bulb-angle S1	Unequal bulb-angle S2	Bulb-tee S3	Lipped channel S4
Centroid G:				
$\bar{x}$	$6.07t$	$4.35t$	–	$5.37t$
$\bar{y}$	$6.07t$	$6.62t$	$5.55t$	–
Shear centre S: e	–	–	–	$6.91t$
Area (A)	$54.9t^2$	$49.9t^2$	$63.9t^2$	$74.0t^2$
Inertia:				
I_{xx}	$2\,605t^4$	$2\,434t^4$	$3\,325t^4$	$12\,371t^4$
I_{yy}	$2\,605t^4$	$1\,239t^4$	$3\,365t^4$	$2\,407t^4$
I_{uu}	$4\,030t^4$	$2\,959t^4$	–	–
I_{vv}	$1\,180t^4$	$725t^4$	–	–
Radius of gyration:				
r_x	$6.89t$	$6.98t$	$7.21t$	$12.93t$
r_y	$6.89t$	$4.98t$	$7.26t$	$5.70t$
r_u	$8.57t$	$7.69t$	–	–
r_v	$4.64t$	$3.81t$	–	–
Torsion constant ($\mathfrak{J}$)	$51.3t^4$	$49.6t^4$	$54.5t^4$	$41.3t^4$
Warping factor (H)	–	–	–	$436\,700t^6$

Notes. 1. The sections are standardized in BS 1161.
2. For section details refer to Figure 9.14.

Table 9.4 Torsional buckling of certain standardized profiles: parameter values

		λ_t	X	s
Single-component members:				
Equal angle	S1	66	0.61	λ_u/λ_t
Unequal angle[2]	S2	57	0.60	$1.4\lambda_u/\lambda_t$
Tee	S3	65	0.78	λ_y/λ_t
Channel	S4	(4)	0.60	λ_x/λ_t
Back-to-back angle members:				
2×S1		70	0.83	λ_y/λ_t
2×S2 (long legs adjacent)		60	0.76	λ_y/λ_t
2×S2 (short legs adjacent)		63	0.89	λ_y/λ_t

Notes. 1. Refer to Figure 9.14 for section geometry.
2. Despite its asymmetry, the value of k for section S2 may be found accurately enough as in Section 9.6.8(4) taking s as given above.
3. For the back-to-back angles it is assumed that the two components are separated by a distance t (at least).
4. For the channel S4: $\lambda_t = 128/\sqrt{1+1660/\lambda_x^2}$

ϕ, θ = parameters to be calculated using the relevant formulae in Table 9.2. Simple values are given for X.

9.6.11 Torsional buckling of certain standardized sections

Figure 9.14 shows four section shapes that are standardized in Britain (BS.1161), having been proposed in 1952 by M. Bridgewater and J.B. Dwight (except for the tee), as efficient profiles with a good compromise between flexural, torsional and local stability. Each comes in a range of geometrically similar sizes, and Table 9.3 lists their section properties (in terms of the thickness t).

Table 9.4 lists seven different shapes of strut section using these profiles, for which the torsional buckling slenderness λ can be readily determined from expression (9.10), with λ_t read direct from the table, while k is found as in 9.6.8(4) using X and s as listed in the table.

9.7 COMBINED AXIAL FORCE AND MOMENT

9.7.1 The problem

When a member has to carry simultaneous axial force P and moment M, it is obviously important to allow for interaction between the two, typical examples being eccentrically loaded axial members and 'beam-columns'. Below we provide checks that cover failure of the cross-section and overall buckling of the member. A simplified procedure is available for checking simple angles, channels and tees connected to one side of a gusset (Section 9.7.9).

9.7.2 Secondary bending in trusses

The members in triangulated truss-type structures, although primarily subject to axial force, also pick up bending moments at their ends due to joint rigidity. These 'secondary' moments can be significant and the question arises as to whether to allow for them in design.

Linear elastic analysis readily enables secondary truss moments to be computed, but the answer obtained is only valid in the early stages of loading. When buckling is imminent, the moments at the ends of a critical member are found to decrease and eventually change sign, changing from disturbing moments to restraining ones. The normal approach for obtaining the calculated static resistance of a truss member is therefore to ignore secondary bending and just consider the axial force. The situation is different in doing a fatigue check, when the secondary bending stresses must be included.

The exception is when there is a significant non-concurrence of the centroidal lines at a node, in which case the resulting eccentricity of loading must always be considered, treated as a case of combined P and M. Normally it is desirable to detail the truss so as to achieve concurrence of the centroidal lines based on the gross section. However, for thin channel-type members with very slender flange elements this is not necessarily valid (Section 9.5.4).

9.7.3 Section classification

As in pure bending, the section must first be classified as fully-compact, semi-compact or slender, unless it is in tension all over. Note that a single classification is needed, corresponding to the particular combination of P and M being applied. Again this is obtained by classifying any individual elements that are wholly or partly in compression, the least favourable element then dictating the classification for the section as a whole.

In considering an element under strain gradient (Section 7.3), the parameter ψ should be based on the usual assumption of flexural behaviour ('plane sections remain plane'). In other words, we put $\psi = y_2/y_1$ where y_1 and y_2 are the distances of the more heavily compressed edge and the other edge from the neutral axis. The assumed neutral axis should be that corresponding to the simultaneous action of P and M, using a plastic stress pattern for the fully-compact or an elastic one for the semi-compact check (Figure 9.15). In either case it is acceptable, for classification purposes, to take this as the axis based on the gross section. When P is high, it is possible for the elastic neutral axis to lie outside the section.

In making the semi-compact check, it is permissible with an 'under-stressed' compression flange to use the same kind of relaxation as that employed in the pure bending case (Section 8.2.8). This consists of

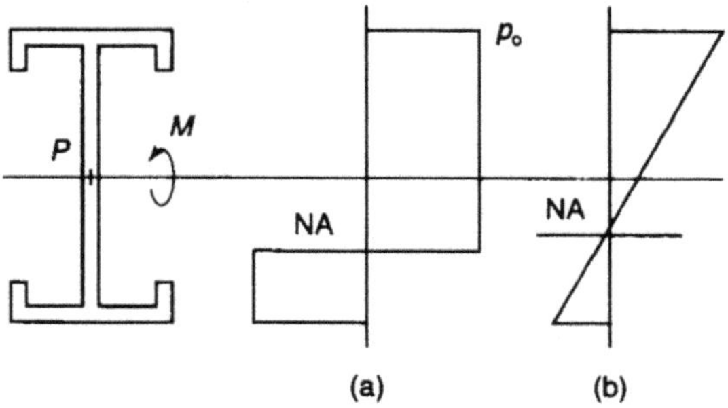

Figure 9.15 Combined axial load and moment, idealized stress patterns: (a) plastic; (b) elastic. NA = neutral axes.

replacing ε in Table 7.1 by a modified value ε' given by:

$$\varepsilon' = \sqrt{\frac{250y_0}{p_o y_c}} \tag{9.17}$$

where y_c = distance from the elastic neutral axis to the element considered, and y_0 = distance from the same axis to the far extreme fibres.

The assumed neutral axis, which again is based on the gross section, should relate to the combined action of P and M.

9.7.4 Interaction formulae (P + uniaxial M)

Below we give four checks (A–D) that apply to members subject to axial load P (applied at the centroid) and moment M, when the axis of M either coincides with an axis of symmetry of the section or is perpendicular to such an axis (Figure 9.16). Table 9.5 indicates which checks are needed for different loading cases.

The checks appear in the form of interaction formulae and are based on BS.8118. They give permissible combinations of P and M, the sign convention being such that every term in each expression is positive. The following general notation is used:

P = axial force arising under factored loading;
M = moment arising at a given cross-section under factored loading;

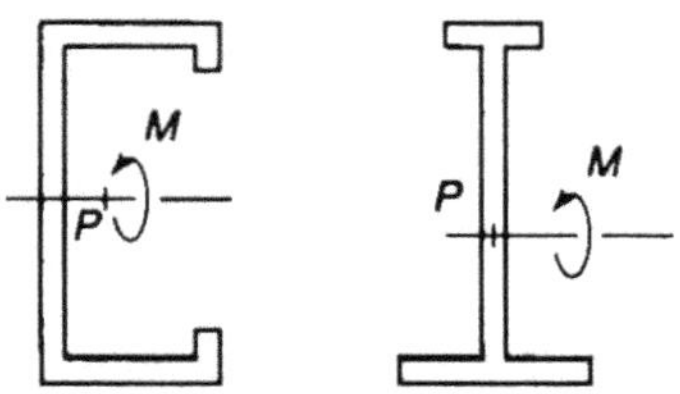

Figure 9.16 Axial load with uniaxial moment.

Table 9.5 Necessary checks for members under combined axial load (P) and moment (M)

Case	P	Axis of M	Checks
1	Tension	Major	A, B, D
2	Tension	Minor	A, B
3	Tension	At θ	E, F, G
4	Compression	Major	A, C, D
5	Compression	Minor	A, C
6	Compression	At θ	E, G

$\overline{M}$ = equivalent uniform value of M, found as in Section 8.7.3;
P_c = calculated resistance to axial load (on its own);
M_c = calculated resistance to moment (on its own).

Check A: Localized failure of the cross-section

$$\frac{P}{P_c} + \frac{M}{M_c} \leqslant \frac{1}{\gamma_m} \tag{9.18}$$

where: P_c = value found as in Section 9.3 based on p_a
M_c = value found as in Section 8.2, reduced to allow for shear.

Check B: General yielding along the length

$$\frac{P}{P_c} + \frac{\overline{M}}{M_c} \leqslant \frac{1}{\gamma_m} \tag{9.19}$$

where: P_c = value found as in Section 9.4 based on p_o
M_c = as for check A, but with no reduction for shear.

Check C: Overall buckling in the plane of the moment

$$\frac{P}{P_c} + \frac{\overline{M}}{M_c} + \frac{P\overline{M}}{2P_c M_c} \leqslant \frac{1}{\gamma_m} \tag{9.20}$$

where: P_c = resistance to overall column buckling in the plane of the moment or to torsional buckling,
M_c = as for check A, but with no reduction for shear.

Check D: Overall buckling out of the plane of the moment

$$\frac{P}{P_c} + \frac{\overline{M}}{Sp_b} \leqslant \frac{1}{\gamma_m} \tag{9.21}$$

where: P_c = resistance to column buckling about other axis,
S = plastic modulus of gross section about axis of M,
p_b = limiting stress for LT buckling (Section 8.7.4).

Note that when P is tensile, the first term in equation (9.21) should be put equal to zero. Also, if the section is of a type not prone to LT buckling, p_b is put equal to M_c/S where M_c is as in check B.

9.7.5 Alternative treatment (P + uniaxial M)

The checks A and B tend to be oversafe when applied to fully-compact profiles, and for these a more realistic result can be obtained by using a direct plastic calculation instead:

$$\text{Modified check A} \qquad M < \frac{S_p p_o}{\gamma_m} \tag{9.22a}$$

$$\text{Modified check B} \qquad \overline{M} < \frac{S_p p_o}{\gamma_m} \tag{9.22b}$$

where p_o is the limiting material stress for bending (Section 5.3), and S_p is a suitably reduced value for the plastic modulus that allows for the presence of P (Section 10.2.3). S_p is different for the two checks.

Figure 9.17 compares results thus obtained with those given by the British Standard rule. It covers check A for a fully-compact box section, and shows how the reduced moment resistance M'_c of the section (in the presence of axial load) varies with P. It is seen that the British Standard rule can underestimate M'_c by as much as 20%.

9.7.6 Interaction formulae (P + biaxial M)

We now consider the case when the applied moment acts about an axis mm inclined at θ to the major principal axis xx (Figure 9.18). Section classification follows the same principles as those given in Section 9.7.3, a single classification being needed which corresponds to the particular

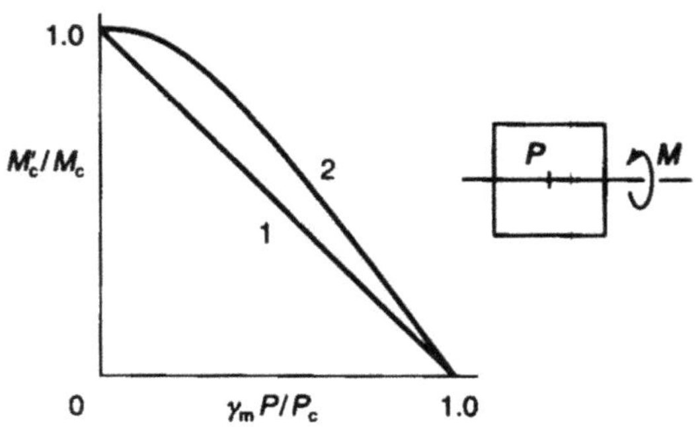

Figure 9.17 Reduced moment resistance M'_c of a fully-compact square box cross-section in the presence of axial load: (1) BS 8118 rule (equation 9.18); (2) Direct plastic calculation (equation (9.22a)).

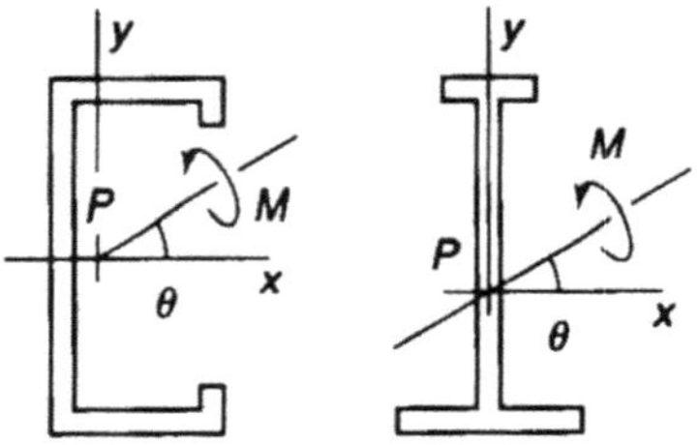

Figure 9.18 Axial load with biaxial moment.

combination of P and M being considered. The relevant checks (E–G) are given below (refer to Table 9.5). P, M and $\overline{M}$ and the sign convention are the same as in Section 9.7.4.

Check E: Localized failure of the cross-section

$$\frac{P}{P_c} + \frac{M\cos\theta}{M_{cx}} + \frac{M\sin\theta}{M_{cy}} \leqslant \frac{1}{\gamma_m} \tag{9.23}$$

where: P_c = value found as in Section 9.3 based on p_a

M_{cx}, M_{cy} = resistance of cross-section to bending about xx and yy, reduced to allow for shear.

Check F: General yielding along the length

$$\frac{P}{P_c} + \frac{\overline{M}\cos\theta}{M_{cx}} + \frac{\overline{M}\sin\theta}{M_{cy}} \leqslant \frac{1}{\gamma_m} \tag{9.24}$$

where: P_c = value found as in Section 9.4 based on p_o

M_{cx}, M_{cy} = as for check A, but with no reduction for shear.

Check G: Overall buckling

$$\frac{\overline{M}\cos\theta}{\overline{M}_{cpx}} + \frac{\overline{M}\sin\theta}{\overline{M}_{cpy}} \leqslant 1 \tag{9.25}$$

where: $\overline{M}_{cpx}$ = permissible value of $\overline{M}$ acting about Gx taken as the lower of the values given by (9.20) and (9.21)

$\overline{M}_{cpy}$ = permissible value of $\overline{M}$ acting about Gy given by (9.20).

The above interaction formulae may also be applied to a skew-symmetric section, if x and y are changed to u, v.

9.7.7 Alternative treatment (P + biaxial M)

The checks E and F sometimes prove unduly pessimistic, in which case better economy may be obtained by using alternative procedures as follows:

1. *Fully-compact sections.* The limiting values of moment for the two checks are found by employing the plastic expressions (9.22) with S_p changed to S_{pm}. Here S_{pm} is a modified value of the plastic modulus, which takes account of the inclination of *mm* as well as the presence of *P* (Section 10.2.3).
2. *Semi-compact and slender sections.* The applied moment is resolved into components $M \cos \theta$ and $M \sin \theta$ about the principal axes, the effects of which are superposed elastically with that of *P*. The section is adequate if at any critical point Q:

Modified check E: $$\left| \frac{P}{p_a A_e} - \frac{yM \cos \theta}{p_o I_{xx}} + \frac{xM \sin \theta}{p_o I_{yy}} < \frac{1}{\gamma_m} \right| \quad (9.26a)$$

Modified check F: $$\left| \frac{P}{A_e} - \frac{y\overline{M} \cos \theta}{I_{xx}} + \frac{x\overline{M} \sin \theta}{I_{yy}} < \frac{p_o}{\gamma_m} \right| \quad (9.26b)$$

where x, y are the coordinates of Q, and the *I*'s relate to the effective section. *P* is taken positive if tensile and negative if compressive. The left-hand side of the equation is always taken positive.

9.7.8 Treatment of local buckling

In applying checks A–D and E–G, care is needed in dealing with elements prone to local buckling. It was explained that the section classification is based on the simultaneous action of *P* and *M* (Section 9.7.3). If on this basis the section comes out as fully or semi-compact, no deduction for local buckling should be made in the determination of P_c and M_c for these checks, even though the section might be regarded as slender under the action of *P* or *M* in isolation. But if (under $P + M$) the section is found to be slender, P_c and M_c should be found exactly as specified in Sections 9.7.4 and 9.7.6, taking (different) effective sections each appropriate to the action considered (*P* or *M*).

9.7.9 Eccentrically connected angles, channels and tees

Axial members of angle, channel or tee-section are often connected to one side of a gusset-plate at their ends. Such members are subject to end moment as well as axial force, because of the eccentricity of loading, and may be rigorously checked using the appropriate interaction equations. Alternatively, an easier method can be adopted as follows without too severe loss of accuracy (based on BS.8118). The procedure is to treat the member as an ordinary tie or strut, and make a simple correction to allow for the eccentricity as follows:

1. *Tension members*. The localized failure check is performed generally as in Section 9.3, but with 60% of any 'outstanding leg' regarded as ineffective. The general yielding check is performed in the usual way with no outstanding leg deduction (Section 9.4).
2. *Compression members*. The check for localized failure will only become critical when the strut is very short, in which case it can be checked as in 1 above. The overall buckling check, which usually governs, consists of ignoring the eccentricity and taking P_c equal to the lower of:

   (a) 40% of the value for column buckling based on centroidal loading, using the radius of gyration r about a centroidal axis parallel to the gusset;
   (b) 100% of the value for torsional buckling (under concentric loading).

This procedure becomes invalid when the attachment to the gusset is inadequate to prevent rotation in the plane of the gusset, as when a single fastener is used.

CHAPTER 10

Calculation of section properties

10.1 SUMMARY OF SECTION PROPERTIES USED

The formulae in Chapters 8 and 9 involve the use of certain geometric properties of the cross-section. In steel design these are easy to find, since the sections used are normally in the form of standard rollings with quoted properties. In aluminium, the position is different because of the use of non-standard extruded profiles, often of complex shape. A further problem is the need to know torsional properties, when considering lateral-torsional (LT) and torsional buckling. The following quantities arise:

(a) *Flexural*	S	Plastic section modulus;
	I	Second moment of area (*inertia*);
	Z	Elastic section modulus;
	r	Radius of gyration.
(b) *Torsional*	$\mathfrak{J}$	St Venant torsion factor;
	I_p	Polar second moment of area (*polar inertia*);
	H	Warping factor;
	β_x	Special LT buckling factor.

The phrase 'second moment of area' gives a precise definition of I and I_p. However, for the sake of brevity we refer to these as 'inertia' and 'polar inertia'.

10.2 PLASTIC SECTION MODULUS

10.2.1 Symmetrical bending

The plastic modulus S relates to moment resistance based on a plastic pattern of stress, with assumed rectangular stress blocks. It is relevant to fully compact sections (equation (8.1)).

Firstly we consider sections on which the moment M acts about an axis of symmetry *ss* (Figure 10.1(a)), which will in this case also be the neutral

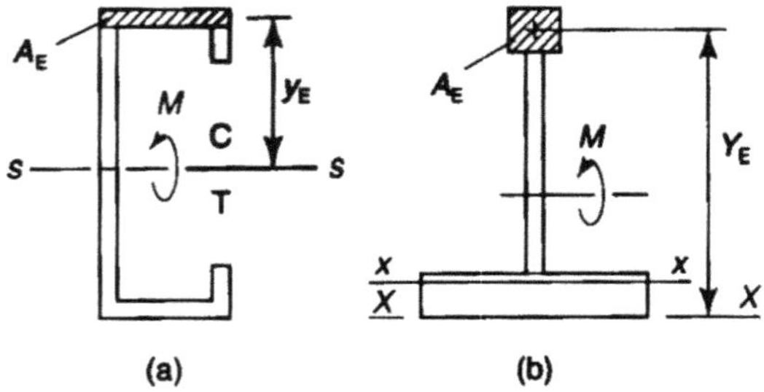

Figure 10.1 Symmetric plastic bending.

axis. The half of the section above *ss* is divided into convenient elements, and *S* then calculated from the expression:

$$S = 2\sum^{C} A_E y_E \tag{10.1}$$

where A_E = area of element, and y_E = distance of element's centroid E above *ss*. The summation is made for the elements lying above *ss* only, i.e. just for the compression material (C).

Figure 10.1(b) shows another case of symmetrical bending, in which *M* acts about an axis perpendicular to the axis of symmetry. The neutral axis (*xx*) will now be the ***equal-area axis***, not necessarily going through the centroid, and is determined by the requirement that the areas above and below *xx* must be the same. *S* is then found by selecting a convenient axis *XX* parallel to *xx*, and making the following summation for all the elements comprising the section:

$$S = \sum A_E Y_E \tag{10.2}$$

where A_E = area of element, taken plus for compression material (above *xx*) and negative for tensile material (below *xx*), and Y_E = distance of element's centroid E from *XX*, taken positive above *XX* and negative below.

The correct answer is obtained whatever position is selected for *XX*, provided the sign convention is obeyed. It is usually convenient to place *XX* at the bottom edge of the section as shown, making Y_E plus for all elements. No element is allowed to straddle the neutral axis *xx*; thus, in the figure, the bottom flange is split into two separate elements, one above and one below *xx*.

10.2.2 Unsymmetrical bending

We now consider the determination of the plastic modulus when the moment *M* acts neither about an axis of symmetry, nor in the plane of such an axis. In such cases, the neutral axis, dividing the compressive material

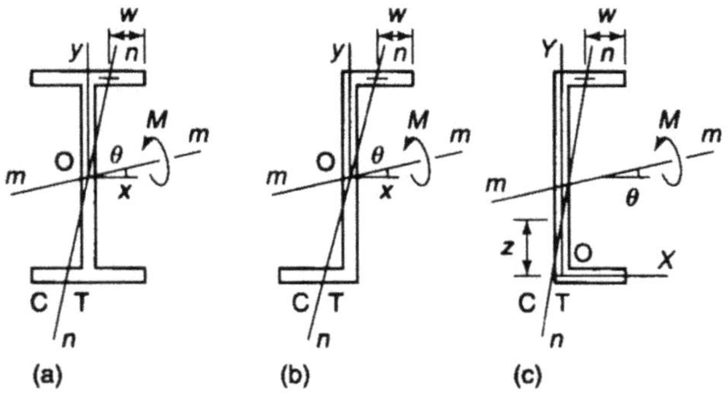

Figure 10.2 Asymmetric plastic bending.

(C) from the tensile material (T), will not be parallel to the axis *mm* about which *M* acts. The object is to determine S_m, the plastic modulus corresponding to a given inclination of *mm*.

Consider first a bisymmetric section having axes of symmetry O*x* and O*y*, with *mm* inclined at θ to O*x* (Figure 10.2(a)). The inclination of the neutral axis *nn* can be specified in terms of a single dimension *w* as shown. Before we can obtain the plastic modulus S_m for bending about *mm*, we must first find the corresponding orientation of *nn* (i.e. determine *w*). To do this, we split up the area on the compression side into convenient elements, and obtain expressions in terms of *w* for the plastic moduli S_x and S_y about O*x* and O*y*, as follows:

$$S_x = 2\sum^{\mathrm{C}} A_{\mathrm{E}} y_{\mathrm{E}} \qquad S_y = 2\sum^{\mathrm{C}} A_{\mathrm{E}} x_{\mathrm{E}} \tag{10.3}$$

where A_E = area of an element, and x_E, y_E = coordinates of the element's centroid E referred to the axes O*x* and O*y*.

In each equation, the actual summation is just performed for the compression material (i.e. for elements lying on one side only of *nn*). The correct inclination of *nn*, corresponding to the known direction of *mm*, is then found from the following requirement (which provides an equation for *w*):

$$S_y = S_x \tan\theta. \tag{10.4}$$

Having thus evaluated *w*, the required modulus S_m is obtained from

$$S_m = \sqrt{(S_x^2 + S_y^2)}. \tag{10.5}$$

The procedure is similar for a skew-symmetric section (Figure 10.2(b)), where a single parameter *w* is again sufficient to define the neutral axis *nn*. In this case, the axes O*x*, O*y* are selected having any convenient orientation. The above equations (10.3)–(10.5) hold good for this case, and may again be used to locate *nn* and evaluate S_m.

We now turn to monosymmetric sections (Figure 10.2(c)). The above treatment holds valid for these in principle, but the calculation is longer, because we need two parameters (w, z) to specify the neutral axis nn. The whole section is split up into convenient elements, none of which must straddle nn. A convenient pair of axes OX and OY is selected, enabling expressions for S_x and S_y (in terms of w and z) to be obtained as follows:

$$S_x = \sum A_E Y_E \qquad S_y = \sum A_E X_E \tag{10.6}$$

where: A_E = area of an element taken positive for compression material and negative for tensile, and X_E, Y_E = coordinates of an elements's centroid E referred to the axes OX and OY, with signs taken accordingly. The summations are performed for the whole section. The dimensions w and z (defining the position of nn) are then obtained by solving a pair of simultaneous equations based on the following requirements:

1. The areas either side of nn must be equal.
2. Equation (10.4) must be satisfied.

Having thus found w and z, the required modulus S_m is determined from expression (10.5) as before. The right answer will be obtained, however the axes OX and OY are chosen, provided all signs are taken correctly.

Unsymmetrical sections can be dealt with using the same kind of approach.

10.2.3 Bending with axial force

The reduced moment capacity of a fully compact section in the presence of a coincident axial force P can be obtained by using a modified value for the plastic section modulus, as required in Section 9.7.5 (expressions (9.22)) and also in Section 9.7.7(1).

Consider first the symmetrical bending case when the moment M acts about an axis of symmetry ss (Figure 10.3(a)). The aim is to find the modified plastic modulus S_p which allows for the presence of P. The figure shows the (idealized) plastic pattern of stress at failure with assumed rectangular stress-blocks, in which we identify regions 1, 2, 3. Region 2 extends equally either side of ss, the regions 1 and 3 being therefore of equal area. Region 2 may be regarded as carrying P, while 1 and 3 look after M. Referring to Section 9.7.5 the stress on region 1 is assumed to be p_a for check A (localized failure), or p_o for check B (general yielding). The height of the neutral axis nn is selected so that the force produced by the stress on region 1 (= its area × p_a or p_o) is equal to $\gamma_m P$. The required value of S_p is then simply taken as the plastic modulus contributed by regions 1 and 3. Note that the value thus obtained is slightly different for the two checks. (P is axial force under *factored* loading.)

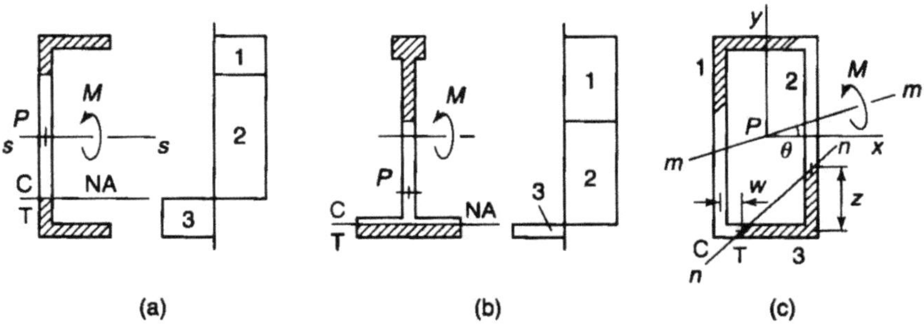

Figure 10.3 Plastic bending with axial load. NA = neutral axis. Hatched areas carry the moment.

A similar approach is used for the other symmetrical case, namely when M acts about an axis perpendicular to ss (Figure 10.3(b)). In defining region 2, it is merely necessary that, apart from having the right area (to carry $\gamma_m P$), it should be so located that regions 1 and 3 are of equal area.

Figure 10.3(c) shows an example of unsymmetrical bending, where the moment M acts about an axis mm inclined at θ to the major principal axis of a bisymmetric section, again in combination with P. In this case, the modified plastic modulus, which we denote by S_{pm}, is a function of P and θ. Two parameters (w, z) are now needed to define the neutral axis nn, as shown, and these may be found from the requirements that: (1) region 2 must have the right area, and (2) the values of S_x and S_y provided by regions 1 and 3 (hatched areas) must satisfy equation (10.4). Having thus located nn, and the extent of the three regions, the required value of S_{pm} is found by using an expression equivalent to equation (10.5).

An identical approach can be used for a skew-symmetric profile, where again two parameters are sufficient to define nn. The same principles apply to other shapes (monosymmetric, asymmetric), but the working becomes more laborious.

10.2.4 Plastic modulus of the effective section

The plastic modulus should when necessary be based on an effective section, rather than the gross one, to allow for HAZ softening at welds and for holes (Section 8.2.4). In considering the HAZ effects, alternative methods 1 and 2 are available (Section 6.6.1).

In method 1, we take a reduced or effective plate thickness $k_{z2}t$ in each nominal HAZ region, instead of the actual thickness, and calculate S accordingly.

Method 2 is convenient for a section just containing longitudinal welds. For bending about an axis of symmetry, it consists of first obtaining S for the gross section, and then deducting an amount $yA_z(1-k_{z2})$ at each HAZ due to the 'lost area' there, where A_z is the nominal softened area and y the distance of its centroid from the neutral axis. For a relatively small longitudinal weld, there is no need to have a very precise value for y. The above approach may also be used when bending is not about an axis of symmetry, except that the lost areas now affect the location of the neutral axis.

10.3 ELASTIC FLEXURAL PROPERTIES

10.3.1 Inertia of a section having an axis of symmetry

The inertia I (second moment of area) must generally be determined about an axis through the centroid G of the section.

To find I_{xx} about an axis of symmetry Gx we split up the area on one side of Gx into convenient elements (Figure 10.4(a)), and use the expression:

$$I_{xx} = 2\sum^{C} I_{Exx} + 2\sum^{C} A_E y_E^2 \tag{10.7}$$

where: I_{Exx} = an element's 'own' inertia about an axis Ex through its centroid E parallel to Gx, A_E = area of the element, and y_E = distance of E from the main axis Gx. The summations are made for the compression material (C) only.

Figure 10.6 shows various common element shapes, for which the value of I_{Exx} and the position of E may be obtained by using the expressions in Table 10.1.

When bending does not occur about an axis of symmetry, but about one perpendicular thereto (Figure 10.4(b)), we have first to locate the neutral axis, i.e. find the position of the centroid G. To do this, we select any convenient preliminary axis XX parallel to the axis of bending. The distance Y_G that the neutral axis lies above OX is given by:

$$Y_G = \frac{\Sigma A_E Y_E}{\Sigma A_E} \tag{10.8}$$

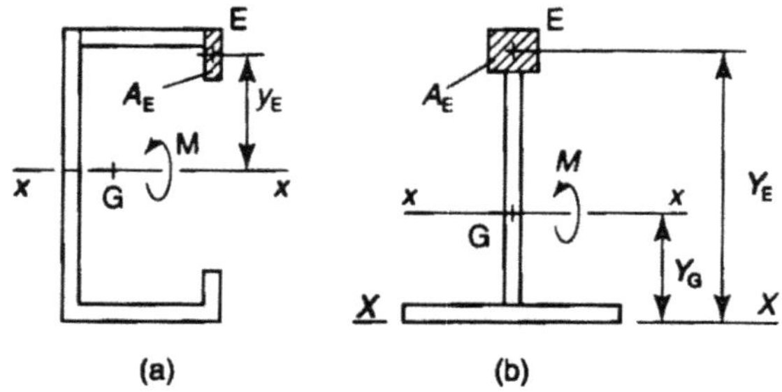

Figure 10.4 Elastic symmetric bending.

where A_E = area of the element (normally taken positive), and Y_E = distance of the element's centroid E from XX taken positive above XX and negative below.

The summations are performed for all the elements. The correct position for the neutral axis is obtained, however XX is chosen, provided the sign of Y_E is taken correctly. A negative value of Y_G would indicate that the neutral axis lies below XX. There are then two possible formulae for I_{xx} which both give the same answer:

Either

$$I_{xx} = \Sigma I_{Exx} + \Sigma A_E(Y_E - Y_G)^2 \tag{10.9}$$

or

$$I_{xx} = \Sigma I_{Exx} + \Sigma A_E Y_E^2 - Y_G^2 \Sigma A_E \tag{10.10}$$

Expression (10.9) is the one taught to students. Expression (10.10) is more convenient for complex sections because a small design change to one element does not invalidate the quantities for all the others.

10.3.2 Inertias for a section with no axis of symmetry

When the section is skew-symmetric or asymmetric (Figure 10.5) we usually need to know the inertias about the principal axes (Gu, Gv), known as the principal inertias (I_{uu}, I_{vv}). The procedure for finding the orientation of the principal axes and the corresponding inertias is as follows:

1. Select convenient preliminary axes Gx and Gy through the centroid G, with Gy directed 90° anti-clockwise from Gx.
2. Calculate the inertia I_{xx} using equation (10.9) or (10.10). Use an equivalent expression to calculate I_{yy}.
3. Calculate the product of inertia I_{xy} (Section 10.3.3).
4. The angle a between the major principal axis Gu and the known axis Gx is then given by:

$$\tan 2a = \frac{2I_{xy}}{I_{xx} - I_{yy}} \tag{10.11}$$

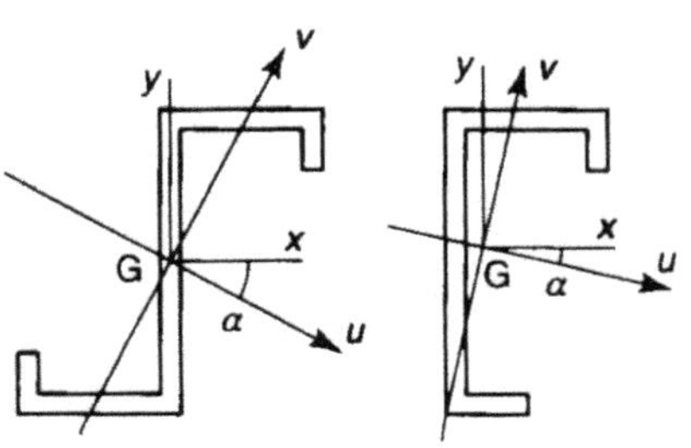

Figure 10.5 Elastic bending, principal axes.

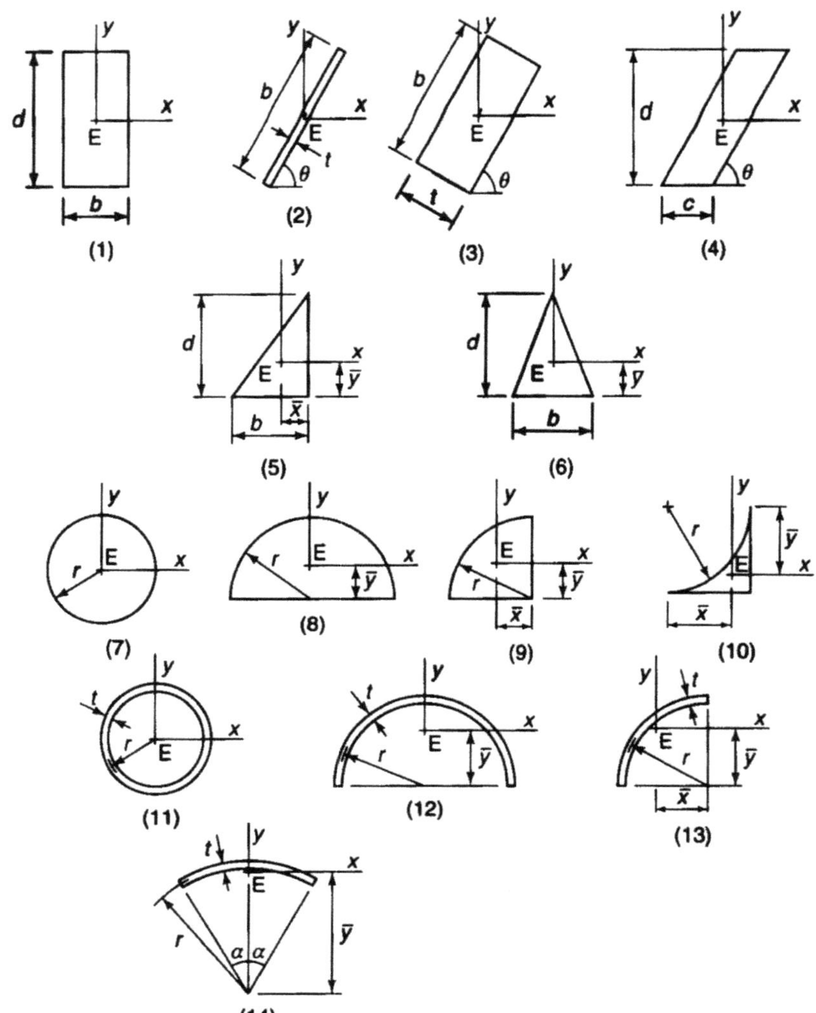

Figure 10.6 Elements of sections.

If α is positive we turn clockwise from Gx to find Gu; if negative we turn anticlockwise. The minor principal axis Gv is drawn perpendicular to Gu.

5. Finally, the major and minor principal inertias are calculated from:

$$I_{uu} = P + \sqrt{(Q^2 + I_{xy}^2)} \qquad I_{vv} = P - \sqrt{(Q^2 + I_{xy}^2)} \tag{10.12}$$

where

$$P = \frac{I_{xx} + I_{yy}}{2} \qquad \text{and} \qquad Q = \frac{I_{xx} - I_{yy}}{2}$$

Table 10.1 Elements of sections – properties referred to axes through element centroid

	Centroid position		Second moment of area ('inertia') about axis through E		Product of inertia
	$\bar{x}$	$\bar{y}$	I_{Exx}	I_{Eyy}	I_{Exy}
1	–	–	$\frac{bd^3}{12}$	$\frac{b^3d}{12}$	0
2	–	–	$\frac{b^3t\sin^2\theta}{12}$	$\frac{b^3t\cos^2\theta}{12}$	$\frac{b^3t\sin 2\theta}{24}$
3	–	–	$\frac{bt(b^2\sin^2\theta+t^2\cos^2\theta)}{12}$	$\frac{bt(b^2\cos^2\theta+t^2\sin^2\theta)}{12}$	$\frac{bt(b^2-t^2)\sin 2\theta}{24}$
4	–	–	$\frac{cd^3}{12}$	$\frac{cd(d^2\cot^2\theta+c^2)}{12}$	$\frac{cd^3\cot\theta}{12}$
5	$\frac{b}{3}$	$\frac{d}{3}$	$\frac{bd^3}{36}$	$\frac{b^3d}{36}$	$\frac{b^2d^2}{72}$
6	–	$\frac{d}{3}$	$\frac{bd^3}{36}$	$\frac{b^3d}{48}$	0
7	–	–	$\frac{\pi r^4}{4}$	$\frac{\pi r^4}{4}$	0
8	–	$\frac{4r}{3\pi}$	$r^4\left(\frac{\pi}{8}-\frac{8}{9\pi}\right)$	$\frac{\pi r^4}{8}$	0
9	$\frac{4r}{3\pi}$	$\frac{4r}{3\pi}$	$r^4\left(\frac{\pi}{16}-\frac{4}{9\pi}\right)$	$r^4\left(\frac{\pi}{16}-\frac{4}{9\pi}\right)$	$r^4\left(\frac{4}{9\pi}-\frac{1}{8}\right)$
10	$0.777r$	$0.777r$	$0.007545r^4$	$0.007545r^4$	$0.004439r^4$
11	–	–	$\pi r^3 t$	$\pi r^3 t$	0
12	–	$\frac{2r}{\pi}$	$r^3t\left(\frac{\pi}{2}-\frac{4}{\pi}\right)$	$\frac{\pi r^3 t}{2}$	0
13	$\frac{2r}{\pi}$	$\frac{2r}{\pi}$	$r^3t\left(\frac{\pi}{4}-\frac{2}{\pi}\right)$	$r^3t\left(\frac{\pi}{4}-\frac{2}{\pi}\right)$	$r^3t\left(\frac{2}{\pi}-\frac{1}{2}\right)$
14	–	$\frac{r\sin\alpha}{\alpha}$	$r^3t\left(\alpha+\frac{\sin 2\alpha}{2}-\frac{2\sin^2\alpha}{\alpha}\right)$	$r^3t\left(\alpha-\frac{\sin 2\alpha}{2}\right)$	0

Notes. 1. Refer to Figure 10.6 for element details.

2. In cases 2 and 11–14 it is assumed that t is small relative to b or r.

10.3.3 Product of inertia

The product of inertia I_{xy}, appearing in Section 10.3.2, can be either positive or negative. It may be calculated using the following expression in which the summations are for all the elements of the section:

$$I_{xy} = \Sigma I_{Exy} + \Sigma A_E x_E y_E \quad (10.13)$$

where I_{Exy} = an element's 'own' product of inertia, referred to parallel axes Ex and Ey through its centroid E, A_E = area of the element (taken positive), and x_E, y_E = coordinates of E referred to the main axes Gx and Gy, with strict attention to signs.

Table 10.1 includes expressions for I_{Exy} for the common element shapes shown in Figure 10.6. In applying these, it is important to realize that I_{Exy} can be either positive or negative, depending on which way round the element is drawn. As drawn in the Figure, it is positive in every case. But if the element is reversed (left to right) or inverted, it becomes negative. If it is reversed and inverted, it becomes positive again.

10.3.4 Inertia of the effective section

The elastic section properties should when necessary be based on an *effective* section, to allow for HAZ softening, local buckling or holes. There are two possible methods for so doing.

In method 1, we take effective thicknesses in the nominal HAZ regions (Section 6.6.1) and effective stress blocks in slender elements (Chapter 7), I being calculated accordingly. The HAZ softening factor k_z is put equal to k_{z2}. If the section is semi-compact, there are no slender elements and thus no deductions to be made for local buckling.

In method 2, we treat the effective section as being composed of all the actual elements that constitute the gross section, on which are then superimposed appropriate negative elements (i.e. 'lost areas'). In any HAZ region, the lost area is $A_z(1 - k_{z2})$. In a slender element it consists of the ineffective material between effective stress blocks (internal elements), or at the toe (outstands). I is obtained basically as in Section 10.3.1 or 10.3.2, combining the effect of all the positive elements (the gross section) with that of the negative ones (the lost areas), and taking A_E, I_{Exx}, I_{Eyy} and I_{Exy} with reversed sign in the case of the latter. In considering the lost area due to softening at a longitudinal weld, there is no need to locate the HAZ's centroid with great precision.

Method 2 tends to be quicker for sections with small longitudinal welds, since it only involves a knowledge of A_z without having to find the actual distribution of HAZ material.

10.3.5 Elastic section modulus

The elastic modulus Z, which relates to moment resistance based on an elastic stress pattern, must always be referred to a principal axis of the section. It is taken as the lesser of the two values I/y_c and I/y_t, where I is the inertia about the axis considered, and y_c and y_t are the perpendicular distances from the extreme compressive and tensile fibres of the section to the same axis. HAZ effects, the presence of holes and local buckling should be allowed for, when necessary, by basing I and hence Z on the effective section.

10.3.6 Radius of gyration

This quantity r is required in calculations for overall buckling. It is simply taken equal to $\sqrt{(I/A)}$ where I is the inertia about the axis considered and A the section area. It is generally based on the gross section, but refer to Section 9.5.4 for sections containing very slender outstands.

10.4 TORSIONAL SECTION PROPERTIES

10.4.1 The torque–twist relation

The torsional section properties $\mathfrak{J}$, I_p, H and β_x are sometimes needed when checking overall member buckling. They should be based on the gross cross-section.

In order to understand the role of $\mathfrak{J}$ and H, it is helpful to consider the simple form of the torque–twist relation for a structural member, when subjected to an axial torque T:

$$T = G\mathfrak{J}\dot{\phi} \tag{10.14}$$

where ϕ = rotation at any cross-section, $\dot{\phi}$ = rate of change of ϕ with respect to distance along the member ('rate of twist'), G = shear modulus of the material, and $\mathfrak{J}$ = St Venant torsion factor.

This equation would be valid if there were no restraint against warping, the idealized condition whereby the torque is applied in a way that leaves the end cross-sections free to distort longitudinally, as illustrated for a twisted I-section member in Figure 10.7(a).

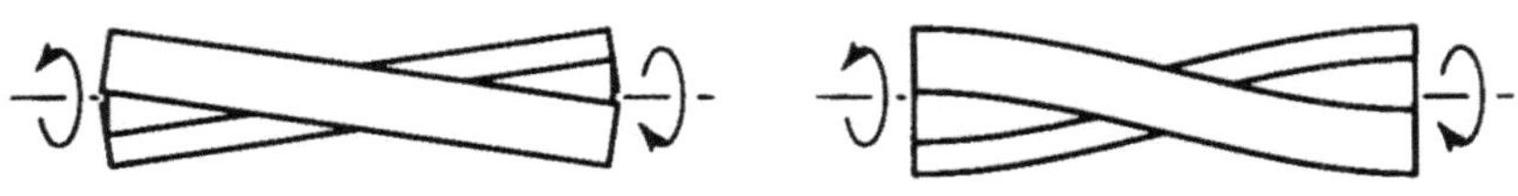

(a) (b)

Figure 10.7 Torsion of an I-beam: (a) ends free to warp; (b) warping prevented.

Usually the conditions are such as to provide some degree of restraint against warping, leading to an increase in the torque needed to produce the same total twist. Figure 10.7(b) shows an extreme case in which the warping at the ends is completely prevented. In any case where warping is restrained, referred to as 'non-uniform torsion', the torque–twist relation becomes:

$$T = G\mathfrak{J}\dot{\phi} - EH\dddot{\phi} \tag{10.15}$$

where $\dddot{\phi}$ = third derivative of ϕ, E = elastic modulus of the material, and H = warping factor.

The quantity $\dddot{\phi}$ is always of opposite sign to $\dot{\phi}$, so that the EH-term in fact represents an increase in T as it should. The section property $\mathfrak{J}$ roughly varies with thickness cubed, whereas H is proportional to thickness. The contribution of the EH-term therefore becomes increasingly important for thin sections.

Equation (10.15) also applies when the torque varies along the length, which is what happens in a beam or strut that is in the process of buckling by torsion. However, 'type-R' sections composed of radiating outstands (Figure 9.7) are unable to warp. For these, H is therefore zero and equation (10.14) is always valid.

10.4.2 Torsion constant, basic calculation

The torsion constant $\mathfrak{J}$ for a typical non-hollow section composed of flat plate elements (Figure 10.8(a)) can be estimated by making the following summation for all of these:

$$\mathfrak{J} = \sum \frac{bt^3}{3} \tag{10.16}$$

where b and t are the width and thickness of an element . For a curved element, it is merely necessary to measure b around the curve. If the thickness t of the element varies with distance s across its width, the

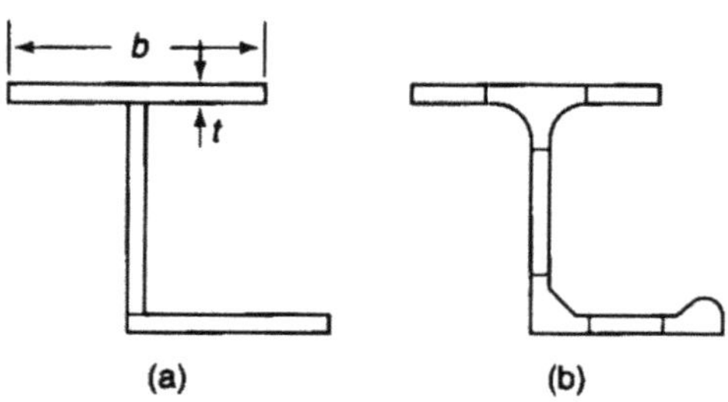

Figure 10.8 Sub-division of the section for finding $\mathfrak{J}$.

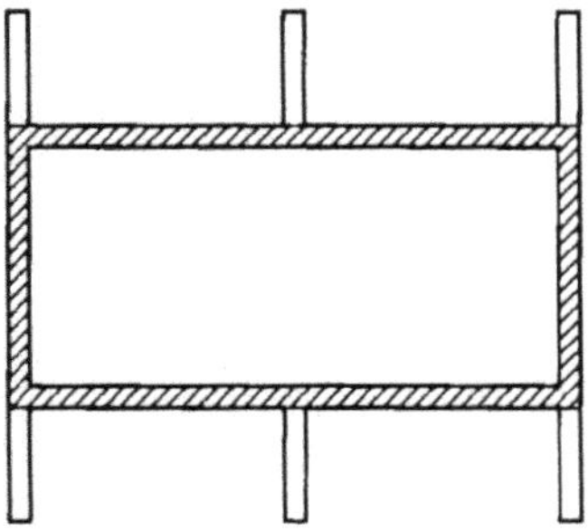

Figure 10.9 Basic torsion box for hollow section.

contribution $\delta\mathfrak{J}$ is taken as:

$$\delta\mathfrak{J} = \int_0^b \frac{t^3\,ds}{3} \tag{10.17}$$

Hollow sections have a vastly increased torsional stiffness, as compared with non-hollow ones of the same overall dimensions. Even so, it is occasionally possible for them to fail torsionally, as when a deep narrow box is used as a laterally unsupported beam. To find $\mathfrak{J}$ for a hollow section (Figure 10.9), we simply consider the basic 'torsion box' (shown shaded) and ignore any protruding outstands. $\mathfrak{J}$ is then given by:

$$\mathfrak{J} = \frac{4A_b^2}{\int \frac{ds}{t}} \tag{10.18}$$

where A_b = area enclosed by the torsion box measured to mid-thickness of the metal, s = distance measured around the periphery of the torsion box, and t = metal thickness at any given position s.

The integral in the bottom line of equation (10.18) is evaluated all around the periphery of the torsion box. For a typical section, in which the torsion box is formed of flat plate elements, it can be simply obtained by summing b/t for all of these.

10.4.3 Torsion constant for sections containing 'lumps'

The torsional stiffness of non-hollow profiles can be much improved by local thickening, in the form of liberal fillet material at the junctions between plates, or bulbs at the toes of outstands. To find $\mathfrak{J}$ for a section thus reinforced we again split it up into elements and sum the contributions $\delta\mathfrak{J}$ (Figure 10.8(b)). For a plate element, $\delta\mathfrak{J}$ is $bt^3/3$ as before; while for a 'lumpy' element, it may be estimated from data given below or else found more accurately using a finite-element program.

Figure 10.10 shows a selection of junction details and also two quasi-bulbs. For any of these the contribution $\delta\mathfrak{J}$ may be found using a

Table 10.2 Torsion constant. Factors used in finding $\delta\mathfrak{J}$ for certain types of section reinforcement

Type	p	q	r	Range	
LJ1	0.99	0.22	0	$N = 1–5$	
LJ2	0.86	0.39	0	$N = 1–5$	
LJ3	0.99	0.22	0	$N = 1–5$	$\alpha = 60–90°$
LJ4	0.86	0.39	0	$N = 1–5$	$\alpha = 60–90°$
LJ5	0.99	0.22	$0.5(T/t - 1)$	$N = 1–4$	$T/t = 1–2$
TJ1	1.03	0.32	0	$N = 1–5$	
TJ2	0.98	0.52	0	$N = 1–5$	
TJ3	1.03	0.32	$0.4(T/t - 1)$	$N = 1–4$	$T/t = 0.5–2$
QB1	0.83	0.39	0	$N = 1–5$	
QB2	0.69	0.20	0	$N = 2–5$	

Notes. 1. The table relates to expression (10.19).
2. Refer to Figure 10.10 for geometry and definition of N.

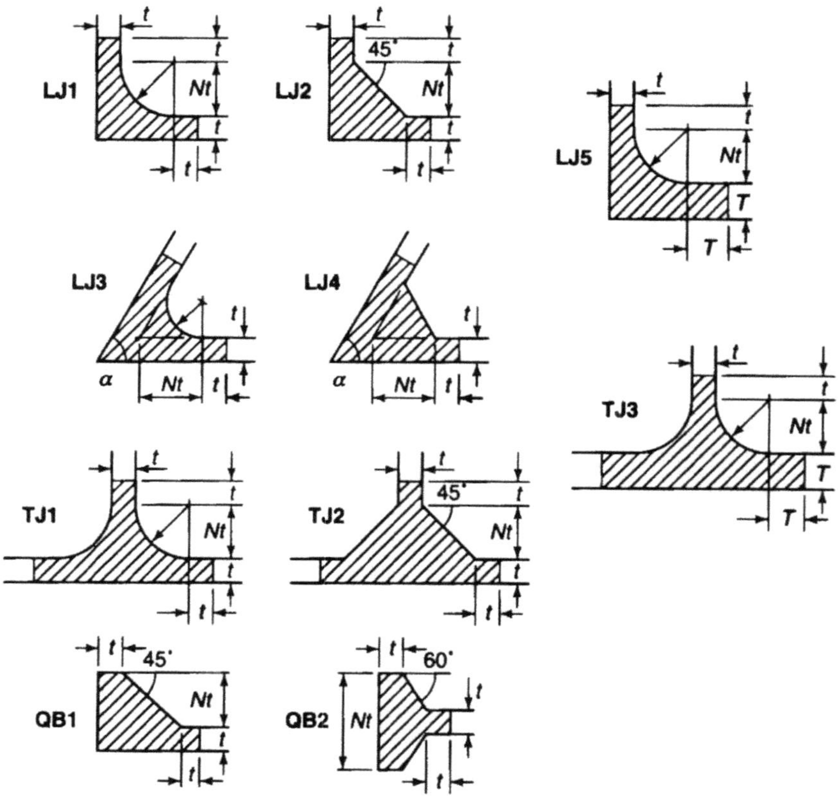

Figure 10.10 Torsion. Section reinforcement covered by expression (10.19) and Table 10.2.

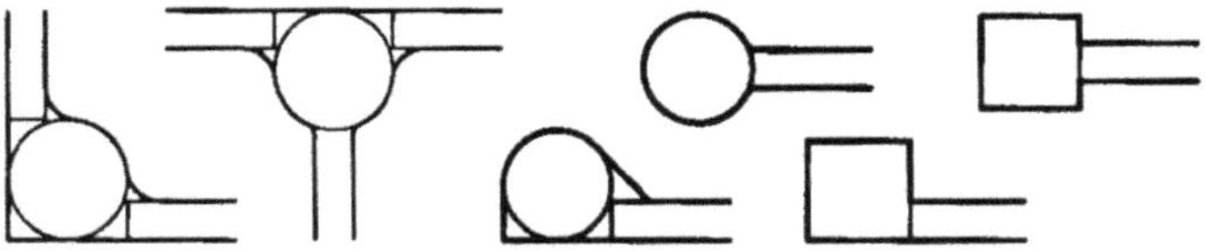

Figure 10.11 Some other forms of section reinforcement.

general expression:

$$\delta\mathfrak{I} = [(p + qN + r)t]^4 \tag{10.19}$$

where N = geometric factor (see Figure 10.10), t = adjacent plate thickness, and p, q, r = quantities given in Table 10.2, valid over the ranges indicated.

Note that for each detail equation (10.19) covers the area shown hatched in Figure 10.10, with the adjacent plate element or elements terminated a distance t or T from the start of the actual reinforcement. This treatment is based on results due to Palmer [28] before the days of computers, using Redshaw's electrical analogue.

Figure 10.11 shows some other kinds of reinforcement, in the form of an added circle or square. For these, it is acceptable to put $\delta\mathfrak{I}$ equal to the simple value below, adjacent plate elements being now taken right up to the circle or square:

$$\text{Circle } \delta\mathfrak{I} = 0.10d^4 \qquad \text{Square } \delta\mathfrak{I} = 0.14a^4 \tag{10.20}$$

where d is the circle diameter, and a the side of the square.

10.4.4 Polar inertia

The polar inertia I_p is defined as the polar second moment of area of the section about its *shear centre* S. S is also the *centre of twist*, the point about which the section rotates under pure torsion. I_p may be calculated as follows:

$$I_p = I_{xx} + I_{yy} + Ag^2 = I_{uu} + I_{vv} + Ag^2 \tag{10.21}$$

where: I_{xx}, I_{yy} = inertias about any orthogonal axes Gx, Gy
I_{uu}, I_{vv} = principal inertias (Section 10.3.2);
A = section area;
g = distance of S from the centroid G.

For sections that are bisymmetric, radial-symmetric or skew-symmetric S coincides with G and $g = 0$. For sections entirely composed of radiating outstand elements, such as angles and tees ('type-R' sections, Figure 9.7), S lies at the point of concurrence of the median lines of the elements. For all others (monosymmetric, asymmetric), a special calculation is generally needed to locate S, involving consideration of warping (Section 10.5). However, in the case of certain shapes, its position may be obtained directly, using simple formulae. These are depicted in Figure 10.12, the

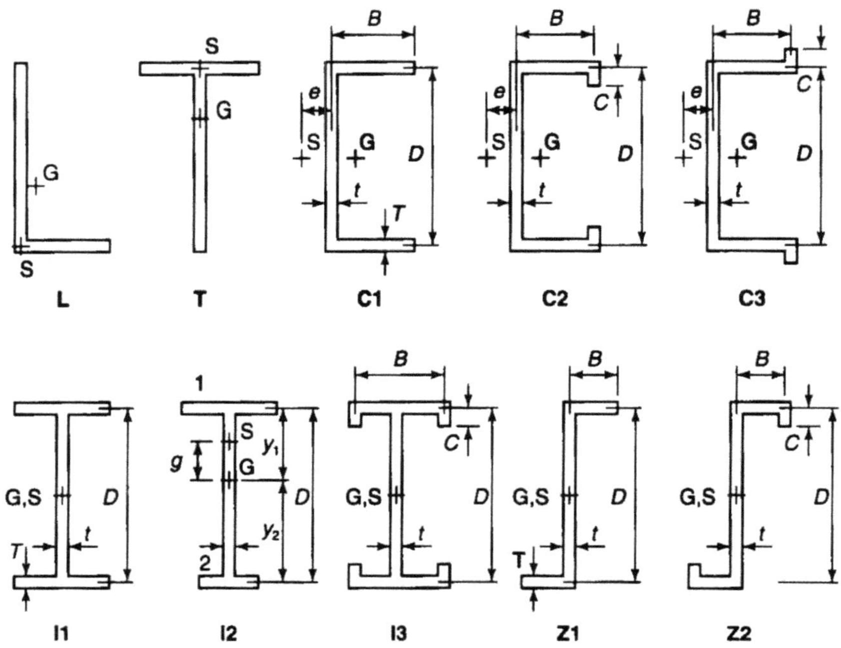

Figure 10.12 Warping. Conventional sections covered by Table 10.3.

corresponding expressions for finding the position of S being given in Table 10.3.

10.4.5 Warping factor

For sections composed of radiating outstands ('type R', Figure 9.7) the warping factor H is effectively zero. For other sections, its determination must generally follow the procedure given in Section 10.5. However, for the shapes covered in Figure 10.12 it may be calculated directly, again using expressions given in Table 10.3.

10.4.6 Special LT buckling factor

The lateral-torsional (LT) buckling of a beam of monosymmetric section, with symmetry about the minor axis (Figure 10.13(a)), involves the *special LT buckling factor* β_x (Section 8.7.5(2)(c)). To obtain this we take axes Gx and Gy as shown, referred to the centroid G. The material on one side only of Gy is split up into convenient elements, none of which may straddle Gy. Then β_x is given by [27]:

$$\beta_x = \frac{2\Sigma j}{I_{xx}} - 2g \qquad (10.22)$$

Table 10.3 Shear-centre position and warping factor for certain conventional sections

Shape	Dimension e, g	Warping factor H
L, T	–	0
C1	$\dfrac{3B}{f+6}$	$\dfrac{D^2B^3T(2f+3)}{12(f+6)}$
C2	$\dfrac{D^2B^2T}{I_{xx}}\left(\dfrac{1}{4}+\dfrac{C}{2B}-\dfrac{2C^3}{3D^2B}\right)$	$\dfrac{B^2T}{6}(D^2B+3D^2C+6DC^2+4C^3)-e^2I_{xx}$
C3	$\dfrac{D^2B^2t}{I_{xx}}\left(\dfrac{1}{4}+\dfrac{C}{2B}-\dfrac{2C^3}{3D^2B}\right)$	$\dfrac{B^2T}{6}(D^2B+3D^2C-6DC^2+4C^3)-e^2I_{xx}$
I1	–	$\dfrac{D^2I_{yy}}{4}$
I2	$\dfrac{y_1I_1-y_2I_2}{I_{yy}}$	$\dfrac{D^2I_1I_2}{I_{yy}}$
I3	–	$\dfrac{D^2I_{yy}}{4}+B^2C^2t\left(\dfrac{D}{2}+\dfrac{C}{3}\right)$
Z1	–	$\dfrac{D^2B^3T(2f+1)}{12(f+2)}$
Z3	–	$\dfrac{B^2t}{12(D+2B+2C)}\{D^2(B^2+2DB+4BC+6DC)+4C^2(3D^2+3DB+4BC+2DC+C^2)\}$

Notes. 1. Refer to Figure 10.12 for section details.
2. For sections C1 and Z1 symbol $f = Dt/BT$.
3. I_{xx} and I_{yy} denote inertia of whole section about xx or yy.
4. For section I2 the symbols I_1 and I_2 denote the inertias of the individual flanges about yy.
5. xx and yy are horizontal and vertical axes through G.

where I_{xx} = inertia about major axis Gx, g = distance of shear centre S from G, and Σj = summation of the quantity j for the elements on the right-hand side of the axis of symmetry.

For any given element, j is defined by the following integral (as given in BS.8118):

$$j = \int y(x^2+y^2)\,dA \tag{10.23a}$$

in which A denotes area. For a thin rectangular element this becomes:

$$j = bty_E(x_E^2+y_E^2+C) \tag{10.23b}$$

where b, t = element width and thickness, and x_E, y_E = coordinates of the midpoint E of the element.

C is obtained as follows depending on the element's orientation:

Parallel to Gx $\quad C = \dfrac{b^2}{12}$

Parallel to Gy $\quad C = \dfrac{b^2}{4}$

Inclined (Figure 10.13(b)) $\quad C = \dfrac{b^2}{12} + \dfrac{Y^2}{6} + \dfrac{XYx_E}{6y_E}$

where X, Y are the projections of the mid-line of the element on either axis.

In applying equations (10.22) and (10.23), the following sign convention must be used, depending on the sense of the applied moment:

1. X and x_E are always taken plus.
2. Axis Gy is directed positive toward the compression side of the section. (The direction selected for Gx is immaterial.)
3. g is taken positive if S lies on the compression side of G, and negative if on the tension side.
4. Y is taken positive if on proceeding along the element y increases with x (as in Figure 10.13(b)), and negative if it decreases.

Note that β_x can be positive or negative, depending on the attitude of the section and the sense of the moment. For a given section in a given attitude the sign of β_x reverses if the applied moment changes from sagging to hogging. Thus for the section in the attitude shown in Figure 10.13(a),

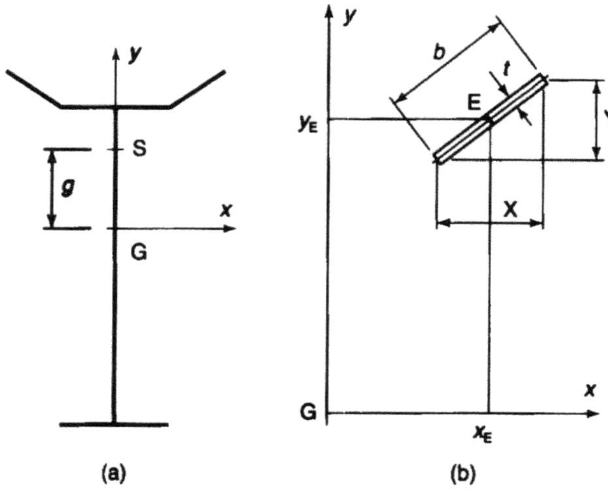

Figure 10.13 Notation used in finding β_x.

the sign of β_x will be plus with a sagging moment and minus with hogging.

10.5 WARPING CALCULATIONS

10.5.1 Coverage

This section gives procedures for locating the shear centre S (when its position is not obvious), and determining the warping factor H, for thin-walled non-hollow profiles composed of plate elements. The warping of such sections is covered in some textbooks, but in a manner that is inconvenient for a designer to use [29]. British Standard BS.8118 does it in a more user-friendly form. Here we present an improved version of the British Standard approach, extended to cover a wider range of section shapes, including those that 'bifurcate'.

10.5.2 Numbering the elements

The plate elements comprising the section should be numbered consecutively (1, 2, 3, 4, etc), starting at a point O and proceeding with a specific *direction of flow* around the section (Figure 10.14). We take O at the point of symmetry when such a point is available and number the elements on one side only of O. In such cases if O comes at the centre of a flat web, we treat this web as two elements either side of O, one of which is numbered. When there is no point of symmetry, we take O at an edge of the section and number all the elements.

In a bifurcating section (Figure 10.14(c)), the two strands beyond the point of bifurcation may be referred to as A and B, with the direction of flow in each always away from the start point O.

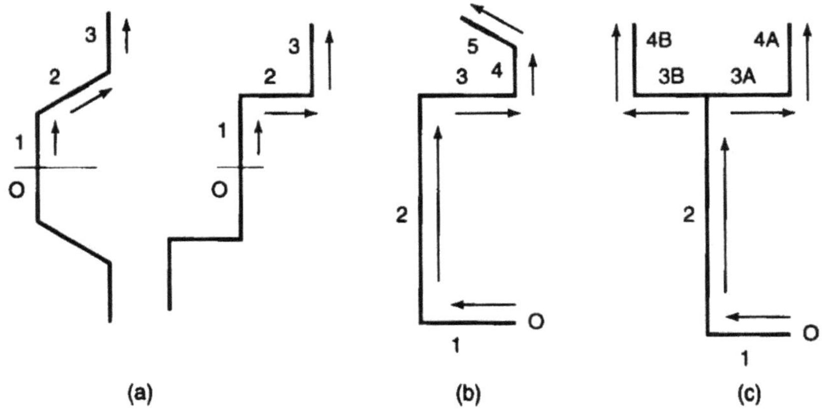

Figure 10.14 Warping: numbering of the elements and 'direction of flow'.

10.5.3 Evaluation of warping

When the ends of a thin-walled twisted member are free to warp (i.e. under no longitudinal restraint, Figure 10.7(a)), each point in the cross-section undergoes a small longitudinal movement z known as the 'warping', which is constant in value down the length of the member. z is measured relative to an appropriate transverse datum plane, and is related to the rate of twist $\dot{\phi}$ by the equation:

$$z = w\dot{\phi} \tag{10.24}$$

in which w, known as the *unit warping*, is purely a function of the section geometry. In order to locate S and determine H, we have to study how w varies around the section.

Figure 10.15 shows an element LN forming part of the cross-section of such a member. The twist is assumed to be in a right-handed spiral (as in Figure 10.7(a)), and the warping movement z (and hence w) is measured positive out of the paper. It can be shown that the increase δw in w, as we proceed across the element in the direction of flow from L to N, is given by:

$$\delta w = bd \tag{10.25}$$

where b = element width, and d = perpendicular distance from the shear-centre S to the median line of the element, taken positive when the direction of flow is clockwise about S, and negative when the flow is anti-clockwise.

For any element r, we define a quantity w_m equal to the value of w at its midpoint M. This is given by:

$$w_m = w_o + w_s \tag{10.26}$$

where w_o = value of w at the start-point O, and w_s = increase in w as we proceed from O to M given by:

$$w_s = \sum_{1}^{r-1} bd + \frac{b_r d_r}{2} \tag{10.27}$$

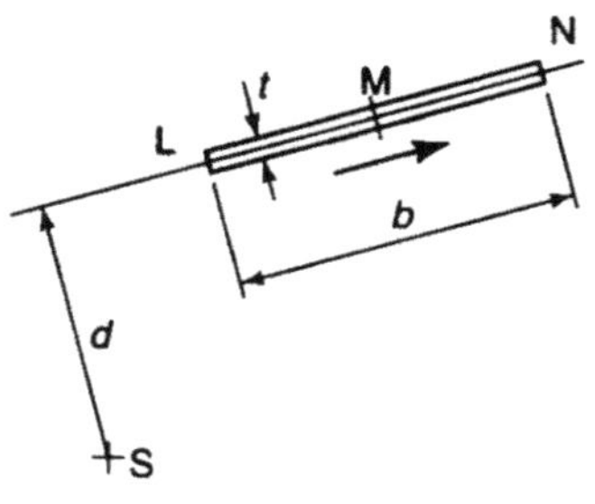

Figure 10.15 Plate element in a member twisted about an axis through the shear-centre S.

The summation in the first term is made for the elements up to, but not including the one considered (element r). In a bifurcating section, it is made for the $(r-1)$ elements on the direct route from O, ignoring side-branches. The first term is zero for the first element.

The value of w_o defines the datum plane, from which z is measured. It can be determined from the requirement that the weighted average of w_m for the whole section should be zero ($\Sigma btw_m = 0$). The value of w_o is therefore given by:

$$w_o = -\frac{1}{A}\Sigma btw_s \qquad (10.28)$$

in which the summation is for all the elements in the section, and A is the section area. For bisymmetric, radial-symmetric and type 1 monosymmetric sections, the start-point O, which is taken at the point of symmetry, is also a point of zero warping, giving $w_o = 0$. In all other cases, w_o is non-zero and must be calculated.

Figure 10.16 compares the warping of an I-section (bisymmetric) and a zed (skew-symmetric). With the zed it is seen that the web, including the point of symmetry O, warps by a uniform non-zero amount.

10.5.4 Formula for the warping factor

The warping factor H can be calculated from the following standard expression, valid for any of the section types covered below:

$$H = \sum bt\left(w_m^2 + \frac{b^2d^2}{12}\right) \qquad (10.29)$$

in which the summation is for all the plate-elements comprising the section. For any element b, t and d are as defined in Figure 10.15, and w_m is found as in Section 10.5.3.

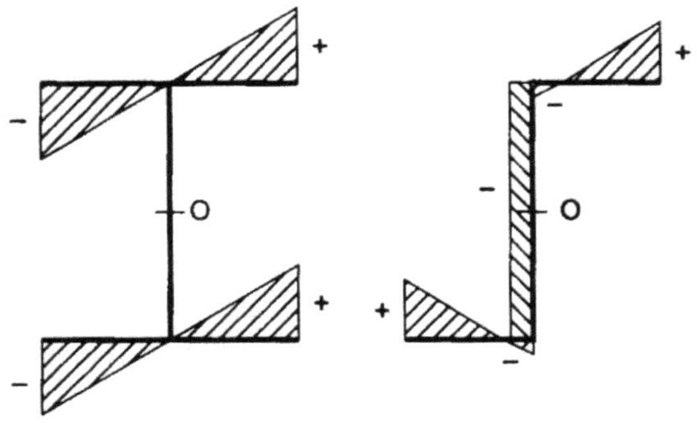

Figure 10.16 Variation of the unit warping w in typical bisymmetric and skew-symmetric sections.

For sections where the position of the shear-centre S is known, equation (10.29) enables H to be found directly. For other sections, it is first necessary to locate S.

10.5.5 Bisymmetric and radial-symmetric sections

For these, S lies at the point of symmetry, which is also the point of zero warping. We take O at the same position, giving $w_o = 0$.

For the *bisymmetric* case (Figure 10.17(a)), the elements are numbered in one quadrant only. The necessary summation in equation (10.29) is then made just for these elements, and the result multiplied by 4 to obtain H. Note that element 1 makes no contribution because it does not warp.

Figure 10.17(b) shows the *radial-symmetric* case, consisting of n equally spaced identical arms, each symmetrical about a radius. For this, the summation in equation (10.29) can be made for the elements lying one side of one arm, and the result then multiplied by $2n$.

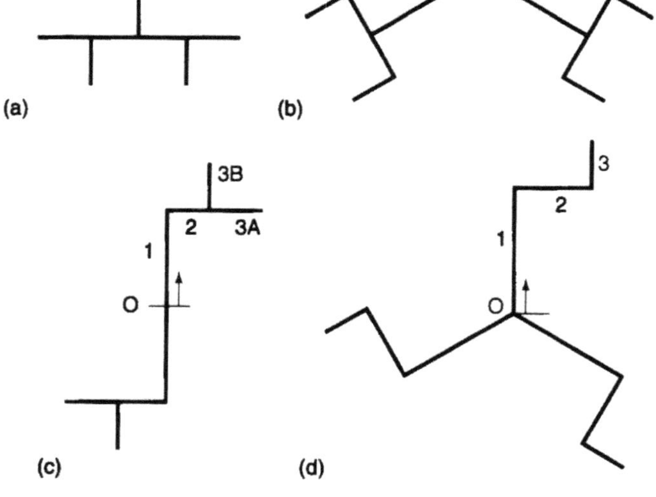

Figure 10.17 Symmetric sections: (a) bisymmetric; (b) radial-symmetric; (c) skew-symmetric; (d) skew-radial-symmetric.

10.5.6 Skew-symmetric sections

Figure 10.17(c) shows a *skew-symmetric* profile. For such a section, S again lies at the point of symmetry, and we take the start-point O for numbering the elements at the same position. But this is not now a point of zero warping and w_o must be evaluated using equation (10.28), enabling w_m to be found for each element. *H* is then calculated using equation (10.29). The summations in equations (10.28) and (10.29) may conveniently be made for the elements in one half of the section (as shown numbered), and then doubled.

Skew-radial-symmetric sections are treated similarly (Figure 10.17(d)), the summations being made for the elements in one arm, and then multiplied by *n* (the number of arms).

10.5.7 Monosymmetric sections, type 1

The shear centre S always lies on the axis of symmetry *ss* for a *monosymmetric* section, but not generally at the same position as the centroid G. Before *H* can be calculated it is necessary to locate S, which involves taking a trial centre-of-twist B at a suitable point on *ss*.

Firstly, we consider monosymmetric sections in which *ss* cuts across the section at its midpoint (Figure 10.18), referred to as type 1. For these, we take B at the point of intersection with *ss*, as shown, which coincides with the start-point O for the numbering of the elements. A quantity w_b is introduced, analogous to w_s, but based on rotation about the trial axis B instead of the true axis S (which has yet to be located). For element *r* we thus have:

$$w_b = \sum_1^{r-1} bd_b + \frac{b_r d_{br}}{2} \tag{10.30}$$

where d_b is the perpendicular distance from B to the median line of an element, using an equivalent sign convention to that employed for *d* (Section 10.5.3). The summation in the first term is for elements lying above *ss*, up to but not including the one considered. The location of S is

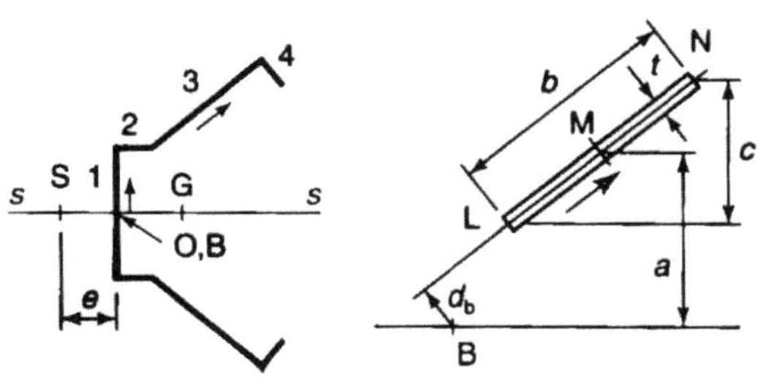

Figure 10.18 Monosymmetric section, type 1.

then given by:

$$e = \frac{2}{I_{ss}} \sum bt\left(aw_b + \frac{cbd_b}{12}\right) \quad (10.31)$$

where: e = distance that S lies to the left of B,
a = distance of midpoint of element from ss (taken positive),
c = projection of b on an axis perpendicular to ss, taken positive if diverging from ss in the direction of flow, and negative if not,
I_{ss} = inertia of the whole section about ss.

The summation is made for the elements lying above ss only.

Having located S it is possible to find d and w_m for the various elements, and hence obtain H from equation (10.29). Distance d is measured from the true axis of rotation (S), and w_m is found from equation (10.26), putting $w_o = 0$. In calculating H the summation in equation (10.29) can be conveniently made for the elements on one side of ss and then doubled.

Alternatively, H may be determined from the following equivalent expression, which avoids the need to find d and w_m for the various elements:

$$H = 2\sum bt\left(w_b^2 + \frac{b^2 d_b^2}{12}\right) - e^2 I_{ss}. \quad (10.32)$$

The summation is again just for the elements lying above ss.

10.5.8 Monosymmetric sections, type 2

We now turn to monosymmetric sections in which the axis of symmetry ss lies along a central web (Figure 10.19), referred to as type 2. These require a modified treatment. The elements are again numbered on one side only

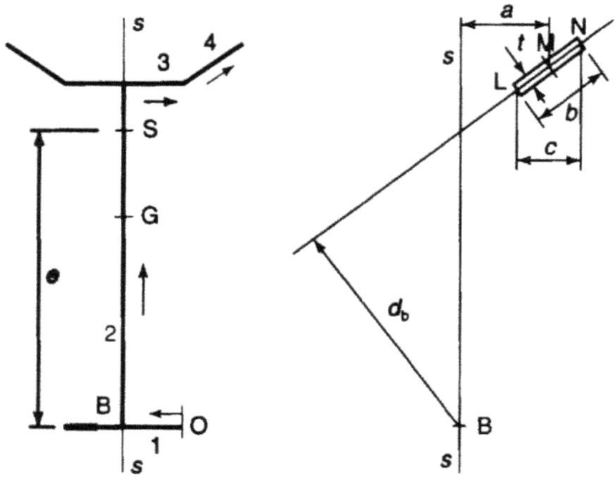

Figure 10.19 Monosymmetric section, type 2.

of *ss*, but from a start-point O now located at an outer edge of the section, as shown, and with the trial axis B taken at the upstream end of the web. A quantity w_{mb} is introduced, analogous to w_m, but based on rotation about B and defined as follows for the element *r*:

$$w_{mb} = w_{ob} + \sum_{1}^{r-1} bd_b + \frac{b_r d_{br}}{2} \tag{10.33}$$

in which d_b is defined as for the type 1 monosymmetric case (Section 10.5.7). The constant term w_{ob} is found from the requirement that the value of w_{mb} for the web must be zero, since this element does not warp, giving:

$$w_{ob} = -\sum_{1}^{s-1} bd_b. \tag{10.34}$$

The summation is for the elements on the direct route from O as far as the web (which is element s), excluding any side-branches.

The distance *e* of the shear-centre S from the trial position B is then given by the following expression:

$$e = \frac{2}{I_{ss}} \sum bt\left(aw_{mb} + \frac{cbd_b}{12}\right) \tag{10.35}$$

in which *a*, *c* and I_{ss} are defined in the same way as for type 1, with the summation made for the elements to the right of *ss* only. A positive value obtained for *e* indicates that S lies above B, and a negative one below.

Having located S, the warping factor *H* can be calculated from the standard expression (10.29), in which the summation can be conveniently made for the elements on one side of ss and then doubled. The quantities *d* and w_m, which relate to rotation about the true axis S, should be found as in Section 10.5.3. In applying equation (10.26) to obtain w_m, we can take w_o as follows:

$$w_o = -\sum_{1}^{s-1} bd \tag{10.36}$$

The summation covers the same elements as for expression (10.34). This correctly brings the warping in the web to zero. Note that, for sections of this type, w_o is more conveniently found thus, rather than using equation (10.28).

Alternatively *H* may be obtained from the following formula, which is equivalent to (10.29) and sometimes more convenient:

$$H = 2\sum bt\left(w_{mb}^2 + \frac{b^2 d_b^2}{12}\right) - e^2 I_{ss} \tag{10.37}$$

The summation is for the elements on one side of *ss* only.

10.5.9 Asymmetric sections

Before starting to find the position of S and the value of H for an *asymmetric* section (Figure 10.20), certain preliminary calculations are needed, namely to locate the centroid G, obtain the orientation of the principal axes Gu and Gv, and determine I_{uu} and I_{vv} (as covered in Section 10.3.2). All the elements of the section must be numbered, starting from O at an outside edge as shown. It does not matter if an element straddles the uu or vv axis.

In order to locate S we introduce a quantity w_g for each element, analogous to w_s, but based on rotation about G. For the rth element, w_g is found as follows (Figure 10.21):

$$w_g = \sum_1^{r-1} bd_g + \frac{b_r d_{gr}}{2} \tag{10.38}$$

where d_g is perpendicular distance from G to the median line of the element, using an equivalent sign convention to that for d in Section 10.5.3. The coordinates U, V of S (referred to axes Gu, Gv) are found as follows, the summations being made for all elements:

$$U = -\frac{1}{I_{uu}} \sum bt\left(a_v w_g + \frac{c_v b d_g}{12}\right) \tag{10.39a}$$

$$V = -\frac{1}{I_{vv}} \sum bt\left(a_u w_g + \frac{c_u b d_g}{12}\right) \tag{10.39b}$$

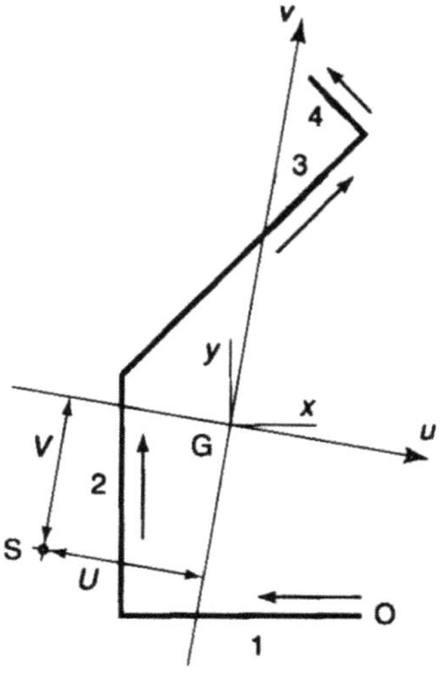

Figure 10.20 Asymmetric section.

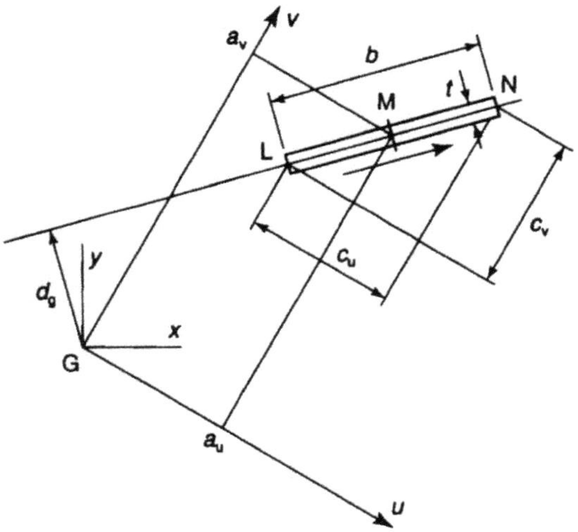

Figure 10.21 Notation for a plate element in an asymmetric section.

where: a_u, a_v = coordinates of the midpoint of an element (referred to Gu, Gv),

c_u, c_v = projected widths of an element on Gu (or Gv), taken positive if u (or v) increases in the direction of flow, and negative if it decreases.

The warping factor H is then found from the standard equation (10.29), in which d and w_m are defined in terms of rotation about S.

CHAPTER 11

Joints

This chapter considers the static strength (ultimate limit state) of aluminium connections, covering joints made with fasteners, welded joints and adhesive-bonded ones. The chapter should be read in conjunction with Chapter 3, which covered the technique of making such joints.

11.1 MECHANICAL JOINTS (NON-TORQUED)

11.1.1 Types of fastener

In this section we look at the strength of aluminium joints made with ordinary fasteners, such as:

- aluminium rivets;
- aluminium bolts;
- stainless steel bolts;
- steel bolts.

British Standard BS.8118 presents a range of materials from which such fasteners might be made. These are listed in Table 11.1, together with suggested limiting stress values for use in design. Other possible aluminium fastener materials are mentioned in Chapter 4 (Section 4.6).

Aluminium rivets can be of conventional solid form. Alternatively, they may be of non-standard proprietary design, especially for use in blind joints (access to one side only). Rivets are not generally suitable for transmitting significant tensile forces.

The bolts considered in this section are 'non-torqued', i.e. they are tightened without specific tension control. For joints loaded in shear, friction between the mating surfaces is ignored and the fasteners are assumed to transmit the load purely by 'dowel action' (shear and bearing). Bolts can be close-fitting or else used in clearance holes, the object of the former being to improve the stiffness of a joint in shear, though not necessarily its strength. Rivets or close-fitting bolts are essential for shear joints in which the transmitted load reverses direction in service, making it necessary to ream out the holes after assembly. When

Table 11.1 Limiting design stresses for selected fastener materials

Fastener		f_u (N/mm²)	Limiting stress p_s (N/mm²)	Limiting stress p_t (N/mm²)	Notes
Aluminium bolts	5056A-H24	310	124	140	$d \leq 12$ mm
	6061-T8	310	124	140	$d \leq 12$ mm
	6082-T6	295	118	133	$d \leq 6$ mm
		310	124	140	$d = 6$–12 mm
					driven
Aluminium rivets	5154A-O, F	215	86	–	Cold or hot
	5154A-H22	245	98	–	Cold
	5056A-O, F	255	102	–	Cold or hot
	5056A-H22	280	112	–	Cold
	6082-T4	200	80	–	Cold
	6082-T6	295	118	–	Cold
Stainless steel bolts (300 series)	A4-50	500	200	275	
	A4-70	700	280	385	
	A4-80	800	320	440	
Steel bolts	grade 4.6	390	156	214	
	grade 6.8	590	236	324	

maximum possible joint stiffness is needed, the designer can resort to friction-grip bolting (Section 11.2).

11.1.2 Basic checking procedure

Possible loading cases are: (a) joints in shear (bolted or riveted); and (b) joints in tension (bolts only). For either of these cases, the basic procedure for checking the limit state of static strength is as follows:

1. Find the greatest transmitted force $\overline{P}$ arising in any one fastener in the group (shear or tension) when factored loading acts on the structure.
2. Obtain the calculated resistance $\overline{P}_c$ for a single fastener of the type used (shear or tension as relevant).
3. The design is satisfactory if:

$$\overline{P} < \frac{\overline{P}_c}{\gamma_m} \tag{11.1}$$

where γ_m is the material factor, which BS.8118 normally takes equal to 1.2 for mechanical joints.

The determination of the fastener force arising (1) follows steel practice and is usually straightforward. The resistance (2) may be obtained using Section 11.1.4 or 11.1.7, although a problem arises here in deciding on a value for the limiting stress. In earlier chapters, we have advocated the use of limiting stresses taken from BS.8118. But, for fasteners, these seem to be inconsistent with other codes, sometimes being very low, and we therefore propose different values.

11.1.3 Joints in shear, fastener force arising

An accurate analysis of a joint in shear, allowing for the true behaviour of the fasteners and the deformation of the connected plates, would be too complex to use in normal design. Instead, the assumption is made that the fasteners respond elastically, while the intervening plate is infinitely stiff, enabling the shear force $\overline{P}$ on any one fastener to be estimated as follows:

1. *Concentric loading on a group of fasteners.* When the transmitted force P arising under factored loading acts through the centroid G of the group (Figure 11.1(a)), it is assumed to be equally shared among all the N fasteners:

$$\overline{P} = \frac{P}{N}. \tag{11.2}$$

2. *Eccentric loading.* When the line of action of P does not go through G (Figure 11.1(b)), it must be resolved into a parallel force P through the centroid and an in-plane moment M as shown. These produce parallel and tangential force components ($\overline{P}_1, \overline{P}_2$) on any given fastener, where:

$$\overline{P}_1 = \frac{P}{N} \tag{11.3}$$

$$\overline{P}_2 = M\left(\frac{r}{\Sigma r^2}\right) \tag{11.4}$$

The summation in equation (11.4) is for all the fasteners in the group, and r the distance of the fastener from G. $\overline{P}$ for the fastener considered is found by combining $\overline{P}_1$ and $\overline{P}_2$ vectorially.

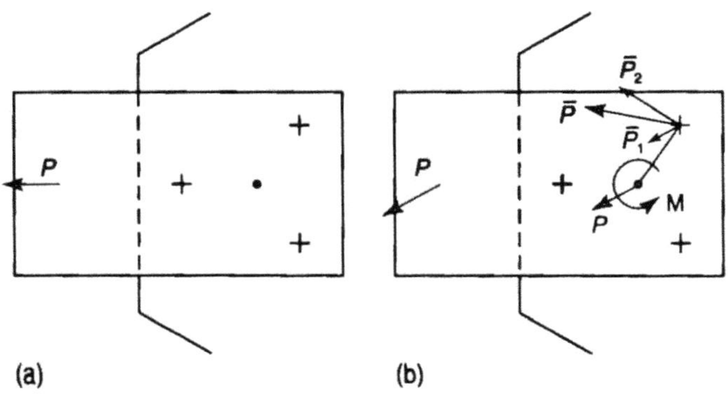

Figure 11.1 Joint in shear under (a) concentric and (b) eccentric loading.

11.1.4 Joints in shear, fastener resistance

Failure at a fastener loaded in shear can occur either in the fastener itself or else in the plate. The calculated resistance $\overline{P}_c$ per fastener should be taken as the lower of two values found as in (1) and (2) below:

1. *Shear failure of the fastener.* This is a relatively sudden form of failure with the fastener shearing into separate pieces. For a conventional fastener the calculated resistance $\overline{P}_c$ is given by:

$$\overline{P}_c = np_s A \tag{11.5}$$

where: p_s = limiting stress in shear (table 11.1) = $0.4f_u$
f_u = minimum ultimate tensile stress of fastener material,
A = shank area (A_1) if failure plane is in shank,
= 'stress-area' (A_2) if it is through the thread (table 11.2),
n = 1 for single-shear joint, 2 for double-shear.

2. *Bearing failure of the plate.* This is a gradual event in which the fastener steadily stretches the hole as the load builds up, there being no clear instant at which failure can be said to have occurred. The calculated resistance $\overline{P}_c$ is taken as follows:

$$\overline{P}_c = kp_p d \sum t \tag{11.6}$$

where: p_p = limiting stress for plate in bearing (table 5.4)
= $1.1(f_{op} + f_{up})$ ... suggested,
f_{op}, f_{up} = 0.2% proof and ultimate stresses for the plate material,
d = shank diameter d_1 if bearing is on shank,
= mean diameter d_2 if it is on the thread (table 11.2),
t = plate thickness,
k = factor depending on the joint geometry (see below),

Table 11.2 Standard ISO bolts (coarse thread)

	Nominal shank d_1 (mm)	Mean diameter d_2 (mm)	Shank area A_1 (mm^2)	Stress area A_2 (mm^2)	Core area A_3 (mm^2)
M3	3	2.6	7.1	5.0	4.1
M4	4	3.7	12.6	8.8	6.1
M5	5	4.4	19.6	14.2	11.8
M6	6	5.3	28.3	20.1	16.1
M8	8	7.1	50.3	36.6	30.2
M10	10	8.9	78.5	58.0	48.5
M12	12	10.8	113	84.3	71.1
M16	16	14.6	201	157	136
M20	20	18.3	314	245	215
M24	24	21.9	452	353	309

and the summation is made for all the plates ('plies') that occur in one or other of the connected components. Values of $\overline{P}_c$ should be obtained for each connected component, using the relevant plate thicknesses, and the least favourable then taken.

The factor k allows for the possibility of the bearing resistance being reduced if: (a) the longitudinal spacing of the holes is too small; or (b) the end hole is too close to the edge of a connected plate in the direction of the transmitted force. With (a), there will be a tendency for the plate to split between holes, and with (b) for the end fastener to tear right through. k should be taken as the lower of the values k_1 and k_2 corresponding to (a) and (b) respectively. Possible values are plotted in Figure 11.2 based on the EU proposals, with BS.8118 included for comparison. Note that Figure 11.2(c) (end distance) ceases to be relevant when the loading is such that the end fastener pushes away from the end of the plate, in which case we take $k_2 = 1.0$. Designs having $s < 2.5d_0$ or $e < 1.5d_0$ are not normally recommended.

Some codes (including BS.8118) require that a bearing check be made for the fastener (as well as for the ply), using a limiting stress based on the fastener material. In common with Canadian and US practice, we reject this as unnecessary.

The above-mentioned treatment takes no account of whether clearance or close-fitting bolts are used. Obviously close-fitting bolts, in holes reamed after assembly, will tend to provide a more uniform distribution of load. But most codes ignore this and adopt the same stresses for clearance bolts as for close-fitting bolts, assuming that the plate metal is ductile enough to even out any differences. The main exception is BS.8118, which allows a 12% higher stress in shear (p_s) for close-fitting bolts and 18% higher for rivets.

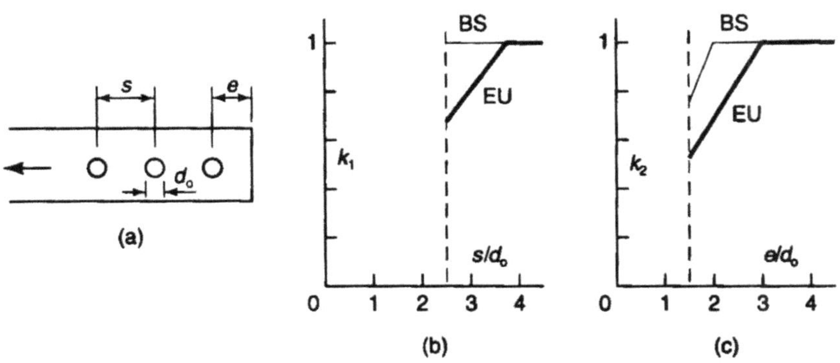

Figure 11.2 Bearing reduction factor: (a) geometry; (b) effect of hole spacing; (c) effect of end distance.

11.1.5 Joints in shear, member failure

A third possible mode of failure in a shear-type joint is tensile failure of the connected member at the minimum net section. This will normally have been covered already under member design. In some situations, however, there is a possibility of joint failure by 'block shear'. The check for this is well covered in Eurocode 3 for steel design, the same principles being applicable to aluminium.

11.1.6 Joints in tension, fastener force arising

Here we consider tension joints made with ordinary bolts or other non-torqued threaded fasteners. For design purposes, any initial tension is ignored, and the joint is analysed as if the bolts were initially done up finger tight. In many joints, the tension $\overline{P}$ arising per bolt under factored loading can be taken as the external force P divided by the number of bolts in the group. In other situations, $\overline{P}$ may be calculated making the same assumptions as in steel.

A problem arises when a connected flange is thin and the bolts are so located as to cause 'prying' action to occur, with a significant increase in the bolt tension. Rules for dealing with this appear in steel codes, but their validity for use with aluminium is not clear.

Although the static design of bolts in tension ignores the initial bolt tension, this does not mean that the tightening of the bolt is unimportant. In all construction it is essential to do bolts up tight: (a) to improve the stiffness of the joint; (b) to prevent fatigue failure of the bolts; and (c) to stop them working loose in service.

11.1.7 Joints in tension, fastener resistance

Here we just consider bolts and other threaded fasteners, rivets being unsuitable for tensile loading. The calculated resistance per fastener may be found from the expression:

$$\overline{P}_c = p_t A_2 \tag{11.7}$$

where: p_t = limiting stress in tension (table 11.1),
= $0.45f_u$ for aluminium bolt,
= $0.55f_u$ for steel or stainless steel bolt,
f_u = minimum ultimate stress of bolt material,
A_2 = 'stress-area' (table 11.2).

The reason for taking a seemingly more conservative p_t-value for aluminium bolts is their lower toughness. And the justification for using the stress area A_2, which is greater than the core area A_3, lies in the redistribution of stress that occurs after initial yielding at the thread root.

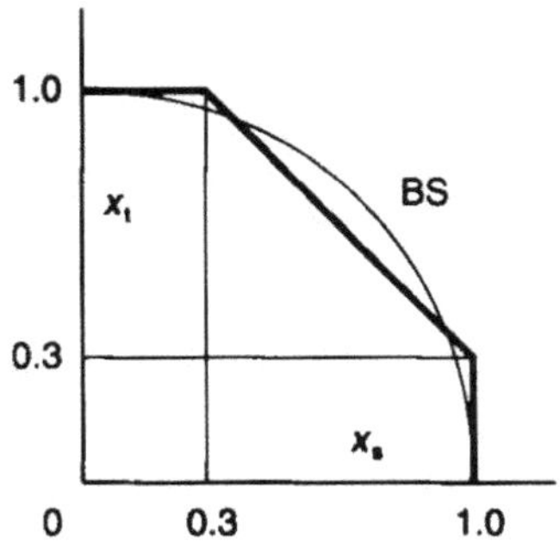

Figure 11.3 Interaction diagram for combined shear and tension on a fastener. $x_s = \gamma_m\overline{P}_s/\overline{P}_{cs}$, $x_y = \gamma_m\overline{P}_t/\overline{P}_{ct}$.

11.1.8 Interaction of shear and tension

When a bolt has to transmit simultaneous shear and tension, one has to consider possible interaction of the two effects. The check for failure of the bolt may be made by using the data plotted in Figure 11.3 in which:

$\overline{P}_s, \overline{P}_t$ = shear and tensile components of force transmitted by any one bolt when factored loading acts on the structure;
$\overline{P}_{cs}, \overline{P}_{ct}$ = calculated resistances to bolt shear failure and bolt tension failure on their own, per bolt;
γ_m = material factor.

The suggested rule, using straight lines, is near enough to the BS.8118 rule, the curve of which is also shown in Figure 11.3. It is more convenient than the latter, in that a designer does not have to bother with an interaction calculation when $\overline{P}_s$ or $\overline{P}_t$ is small.

The check for bearing failure of the ply is performed in the usual way with the bolt tension ignored.

11.1.9 Comparisons

Our suggested limiting stresses for a selection of materials are given in Tables 11.1 (fastener material) and 5.4 (plate material). These values, which are based on the expressions in Sections 11.1.4 and 11.1.7, differ from those in BS.8118. The reason for not following the British Standard is that the stress values it employs seem inconsistent, and sometimes rather low when compared with other codes. Table 11.3 compares the various expressions used in the two treatments, while Table 11.4 lists some actual stress values calculated for typical materials. The following points affect these comparisons:

Table 11.3 Comparisons with BS.8118 – expressions for the limiting fastener stress

			BS.8118	Suggested
Shear	Aluminium bolt	$p_s =$	$0.286 f_u$	$0.4 f_u$
	Stainless steel bolt		$0.714 f_o$ or $0.3(f_o + f_u)$	$0.4 f_u$
	Steel bolt		$0.595 f_o$	$0.4 f_u$
Bearing	Aluminium bolt	$p_p =$	$1.12 f_u$	Ignore
	Stainless steel bolt		$2.4 f_o$ or $(f_o + f_u)$	Ignore
	Steel bolt		$2 f_o$	Ignore
	Ply		$2.4 f_{op}$ or $(f_{op} + f_{up})$	$1.1(f_{op} + f_{up})$
Tension	Aluminium bolt	$p_t =$	$0.336 f_u$	$0.45 f_u$
	Stainless steel bolt		$1.2 f_o$ or $0.5(f_o + f_u)$	$0.55 f_u$
	Steel bolt		f_o	$0.55 f_u$

Notes. 1. f_o, f_u = proof (yield) and ultimate stress of bolt material.
f_{op}, f_{up} = the same for ply material.
2. For shear the bolts are assumed to be in normal clearance holes.
3. Where two values are listed, the lower is taken (BS.8118).

Table 11.4 Comparisons with BS.8118 – typical limiting stresses for fasteners

			BS.8118 (N/mm^2)	Suggested (N/mm^2)
Shear	Aluminium bolt	$p_s =$	89	124
	Stainless steel bolt	$p_s =$	150	200
	Steel bolt	$p_s =$	143	156
Bearing	Aluminium bolt	$p_p =$	347	Ignore
	Stainless steel bolt	$p_p =$	504	Ignore
	Steel bolt	$p_p =$	480	Ignore
	Ply	$kp_p =$	(a) 555 (b) 555	(a) 407 (b) 610
Tension:	Aluminium bolt	$p_t =$	104	140
	Stainless steel bolt	$p_t =$	252	275
	Steel bolt	$p_t =$	240	215

Notes. 1. Materials covered:

Aluminium bolt	(6082-T6)	$f_o = 270$	$f_u = 310$ N/mm^2
Stainless steel bolt	(A4-50)	210	500 N/mm^2
Steel bolt	(Grade 4.6)	240	390 N/mm^2
Ply – 6082-T6 plate	($t > 6$ mm)	255	300 N/mm^2

2. Bearing on ply. The stresses have been factored by k in order to allow for the joint geometry (Figure 11.2), assuming: (a) $s = 3d_0$, $e = 2d_0$; (b) $s = 4d_0$, $e = 3d_0$ where s = longitudinal pitch, e = edge distance, d_0 = hole diameter.

1. *Shear failure of fastener*. The expressions for p_s in Table 11.3 refer to bolts used in normal clearance holes. British Standard BS.8118 allows a 12% higher value for close-fitting bolts, and 18% higher for cold-driven rivets. Our proposals, like other codes, make no distinction.
2. *Bearing failure of plate*. The two treatments effectively take different values for the joint-geometry factor k (Figure 11.2). In Table 11.4, the listed stresses for ply bearing have been multiplied by the relevant k in

order to give a fair comparison, two different geometries being thus covered.

Our suggested treatment is broadly tailored to be slightly conservative when compared with what has been proposed for the European (EU) draft, when due account is taken of the different load factors used (γ values). We differ from the EU draft in our approach to bearing. First, we only require a designer to check bearing on the ply, as in USA and Canada, and ignore bearing on the fastener. Secondly, our expression for p_p includes both the proof and the ultimate stress of the ply material, whereas the draft Eurocode relates it only to the ultimate. Our justification for this is that ply-bearing concerns the gradual stretching of the hole, which must be a function of the proof as well as the ultimate.

It is seen that the BS.8118 values for aluminium bolts in shear and in tension are very low when compared with our suggested treatment. For bearing on the ply, it is remarkable that the British Standard value is 50% higher than that for bearing on an aluminium bolt made of the same material.

11.1.10 Joints made with proprietary fasteners

When joints are made using special rivets of 'proprietary' design (Section 3.2.3), designers will probably rely on the manufacturer of these for strength data. Alternatively, they may conduct their own tests to establish the resistance. In either case it is necessary to consider carefully the value to take for γ_m, and a value higher than the usual 1.2 might be appropriate.

11.2 MECHANICAL JOINTS (FRICTION-GRIP)

11.2.1 General description

High-strength friction-grip (HSFG) bolts, made of high tensile steel, are employed for joints loaded in shear when joint stiffness at working load is the prime requirement. They are used in clearance holes and, until slip occurs, transmit the load purely by friction between the plate surfaces. Under service loading, they thus provide a rock-solid connection, much stiffer than when close-fitting conventional fasteners are used (non-torqued). The bolts are made of high-strength steel and are torqued up to a high initial tension, so as to generate enough friction to stop slip occurring at working load. The control of tightening and the preparation of the plate surfaces is critical (Section 3.2.2).

High-strength friction-grip bolts were developed in USA in the 1940s for use on steel, following pre-war work by Professor C. Batho at Birmingham University in Britain. They are less attractive for use with aluminium because:

1. The coefficient of friction (slip-factor) is less.
2. When the connected plates are stressed in tension, the decrease in bolt tension and hence in friction capacity is more pronounced.
3. The bolt tension falls off with decrease in temperature.
4. HSFG bolts are not suitable for use with the weaker alloys. (BS.8118 bans their use when the 0.2% proof stress of the plates is less than 230 N/mm^2.)

High-strength friction-grip joints must be checked for the ultimate limit state, as with non-torqued fasteners, and also for the serviceability limit state. The latter check is needed to ensure that gross slip, and hence sudden loss of joint rigidity, does not occur in service.

11.2.2 Bolt material

British Standard BS.8118 states that only general grade HSFG bolts should be used with aluminium, the specified minimum properties of which are as follows (BS.4395: Part 1):

Stress at permanent set limit	635 N/mm^2
Tensile strength (ultimate stress)	825 N/mm^2

11.2.3 Ultimate limit state (shear loading)

It is assumed for this limit state that gross slip has occurred, and that any residual friction is not to be relied on. All the hole clearance is taken up and the load is transmitted by dowel action, in the same way as for conventional (non-torqued) bolts. The joint should therefore be checked as in Section 11.1.2. In so doing, the transmitted shear force $\overline{P}$ per bolt arising under factored loading is found as in Section 11.1.3; and the calculated resistance $\overline{P}_c$ as in Section 11.1.4(2). There is unlikely to be any need to check for shearing of the bolt, because of the very high strength of its material.

11.2.4 Serviceability limit state (shear loading)

The checking of HSFG bolts for this limit state proceeds basically in the following manner:

1. Find the greatest transmitted shear force $\overline{P}_n$ arising in any one bolt, when nominal (unfactored) loading acts on the structure.
2. Obtain the calculated friction capacity $\overline{P}_f$ per bolt.

3. The design is satisfactory if for any one bolt

$$\overline{P}_n < \frac{\overline{P}_f}{\gamma_s} \tag{11.8}$$

where γ_s is the serviceability factor (Section 11.2.7).

The shear load arising per bolt ($\overline{P}_n$) is determined using the same kind of calculation as that employed with non-torqued fasteners (Section 11.1.3), but taking nominal instead of factored loading. This implies that there is microscopic slip at the various bolts, as redistribution of load takes place. Such microslip or 'creaking' under service conditions must not be confused with the gross slip that occurs when the friction finally breaks down.

The calculated friction capacity ($\overline{P}_f$) depends on the reaction force $\overline{R}$ between the plate surfaces (per bolt) and on their condition. It is given by the expression:

$$\overline{P}_f = n\mu\overline{R} \tag{11.9}$$

where n = number of friction interfaces and, μ = slip-factor (Section 11.2.6).

11.2.5 Bolt tension and reaction force

A fabricator installing HSFG bolts is required to use a torquing procedure which ensures that the initial tension in the as-torqued condition is not less than the specified proof load T_o for the size being used. Here T_o is the tension corresponding to a stress, calculated on the stress-area A_2 of the bolt, that is slightly below the minimum permanent set value of the bolt material. For general grade HSFG bolts to BS.4395: Part 1 this stress is 590 N/mm^2, and Table 11.5 gives the resulting T_o-values for a range of bolt sizes.

Ideally, a designer would hope to take the reaction force $\overline{R}$ in equation (11.9) equal to the bolt proof load T_o. In practice, there are four possible situations in which $\overline{R}$ becomes reduced:

1. The joint has to transmit an external tensile load, acting in the axial direction of the bolts, as well as the shear load.
2. The connected plates are under in-plane tension (Poisson's ratio effect).

Table 11.5 Proof load T_o for HSFG bolts

	M12	M16	M20	M22	M24
Stress area A_2 (mm^2)	84	157	245	303	353
Proof load T_o (kN)	49	92	144	177	207

Note. The quoted values apply to general grade HSFG bolts (BS.4395: Part 1) and are based on a stress of 590 N/mm^2 acting on the stress area.

3. The operating temperature is significantly lower than when the bolts were tightened (differential thermal contraction).
4. There is poor fit-up because the plates are initially warped.

In situation (1) it is essential to make a suitable calculation and obtain a modified value for $\overline{R}$. This may be done using the following expression:

$$\overline{R} = T_o - kT_1 \tag{11.10}$$

in which T_1 is the applied external tensile force that arises under nominal (unfactored) loading per bolt. The factor k is typically taken as 0.9 in steel design. In aluminium construction, because of the lower modulus of the aluminium plates, a larger proportion of T_1 is used up in increasing the bolt tension, and the drop in $\overline{R}$ is correspondingly less. Despite this, BS.8118 still takes $k = 0.9$ in equation (11.10). We would recommend $k = 0.8$ as a reasonable and more favourable design value.

The Poisson's ratio effect when the connected plates are stressed in tension (situation (2)) is more of a factor with aluminium than it is with steel, because of the lower modulus E and the higher Poisson's ratio. In steel, the effect is usually ignored. In aluminium, the reduction in $\overline{R}$ is greater, possibly reaching 10 or 20% of T_o under the worst conditions. The designer must therefore decide whether or not to make an arbitrary adjustment to $\overline{R}$, depending on the level of the tensile stress in the plates. British Standard BS.8118 suggests that no allowance is necessary until the plate stress under nominal loading reaches 60% of the proof stress.

The temperature effect (3) does not arise in an all-steel joint. In aluminium, if the ambient temperature falls to 30°C below that at the time of torquing, one would expect a decrease in $\overline{R}$ of about 10% of T_o due to this effect. Again it is up to the designer to decide whether any allowance is needed, depending on the environment.

Poor fit-up (4) only becomes a factor in massive joints between thick plates containing a lot of bolts. It can be minimized by employing a careful torquing sequence, and is likely to be less serious in aluminium anyway because of the lower modulus.

11.2.6 Slip factor

The value to be taken for the slip factor μ in equation (11.9) depends critically on the preparation of the plate surfaces before assembly. British Standard BS.8118: Part 1 states that if a standard procedure is used, which involves blasting with G38 grit (refer to BS.2451), the value $\mu = 0.33$ may be taken, compared with a typical figure of 0.45 used in steel design. Note however that there seems to be inconsistency between Parts 1 and 2 of BS.8118 in this respect.

When the method of surface preparation does not follow the standard procedure, μ should be based on tests, which may be carried out in accordance with BS.4604: Part 1. The resulting μ-value may be lower or higher than 0.33, depending on the procedure used.

11.2.7 Serviceability factor

British Standard BS.8118 is confusing as to the value to be used for the serviceability factor γ_s in equation (11.8). When the slip-factor μ is taken as 0.33, based on the standard surface preparation, alternative values $\gamma_s = 1.2$ and 1.33 are given and it is not clear which to use when. On the other hand, when μ is found from tests, BS.8118 permits $\gamma_s = 1.1$. An appropriate value would lie in the range 1.1–1.2. The value of 1.33 seems too high.

11.3 WELDED JOINTS

11.3.1 General description

In this section, we consider the static strength (ultimate limit state) of aluminium welds made by the MIG or TIG process. There are two essential differences from steel: firstly, the weld metal is often much weaker than the parent metal; and, secondly, failure may occur in the heat-affected zone (HAZ) rather than in the weld itself. The weld metal can be stronger or weaker than the HAZ material, depending on the parent/filler combination. It tends to be less ductile and better joint ductility is obtained when failure occurs in the HAZ. Refer to Chapter 6 for data on HAZ softening.

The various action-effects that a weld may be required to transmit, possibly occuring in combination, are as follows (Figure 11.4):

(*a*) *transverse force* acting perpendicular to the axis of the weld;
(*b*) *longitudinal force* acting parallel to the weld axis;
(*c*) *axial moment* acting about the weld axis.

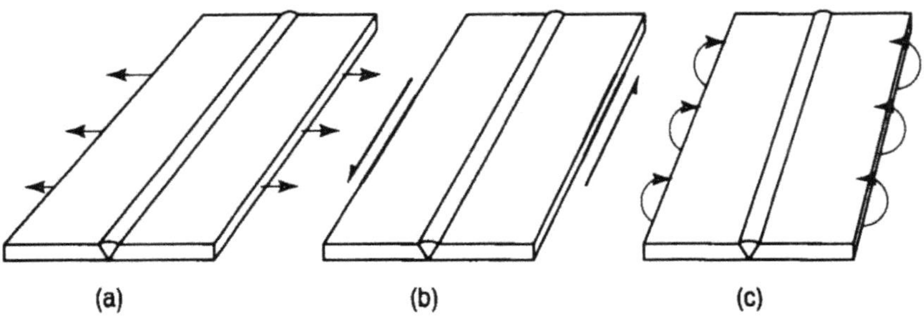

Figure 11.4 Action-effects at a weld.

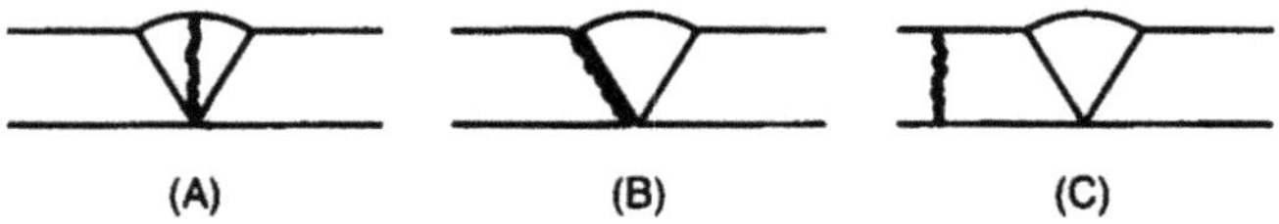

Figure 11.5 Weld failure planes: (A) weld metal; (B) fusion boundary; (C) HAZ.

For each of these there are three possible planes on which failure can occur (Figure 11.5):

A. Joint failure in the *weld metal*;
B. Joint failure in the HAZ at the edge of the weld deposit, known as *fusion boundary* failure;
C. Failure in the HAZ at a small distance from the actual weld. This mainly applies to heat-treated 7xxx-series material for which there is a dip in the HAZ properties at the position concerned (Figure 6.2).

The resistance must be checked on all three planes, except with a full penetration butt under transverse compression for which it is only necessary to consider plane C. Note that failure on plane C has already been covered under member design in Chapters 8 and 9. Here we just consider A and B.

11.3.2 Basic checking procedure

Looking first at the loading cases (a) and (b), the basic procedure for checking a weld against failure on plane A or B is as follows:

1. Find the force $\overline{P}$ (transverse or longitudinal) transmitted per unit length of weld, when factored loading acts on the structure (Section 11.3.3).
2. Obtain values of the calculated resistance $\overline{P}_c$ per unit length of weld (transverse or longitudinal as relevant), corresponding to weld metal failure (Section 11.3.4) and fusion boundary failure (Section 11.3.5).
3. The weld is acceptable if, at any position along its length, the following is satisfied for both the possible failure planes:

$$\overline{P} < \frac{\overline{P}_c}{\gamma_m} \tag{11.11}$$

where γ_m is the material factor (Section 5.1.3). Weld strength is notoriously difficult to predict, and a designer should resist the temptation to use too low a value for γ_m. British Standard BS.8118 requires it to be taken in the range 1.3–1.6 for welded joints, depending on the level of control exercized in fabrication. The value $\gamma_m = 1.3$ is permitted when the procedure meets normal quality welding, as laid down in Part 2 of the BS.8118 (Section 3.3.5).

11.3.3 Weld force arising

In many cases, the intensity $\overline{P}$ of the transmitted force per unit length of weld may be simply taken as force divided by length:

$$\overline{P} = \frac{P}{L_e} \tag{11.12}$$

where P = total transmitted force (either transverse or longitudinal) when factored loading acts on the structure, and L_e = effective length of weld.

The effective length L_e would normally be taken as the actual weld length L less a deduction at either end equal to the weld throat, to allow for possible cracks or craters. Such deduction need not be made if end defects are avoided by the use of run-on and run-off plates during fabrication, or by continuing a fillet weld around the corner at an end.

Equation (11.12) is valid when the loading on a weld is uniform. A problem arises when longitudinal fillet welds are used to transmit the load at the end of a tension or compression member. With these, elastic analysis predicts that the intensity of the transmitted force (even though nominally uniform) will peak at either end of the weld, and the use of expression (11.12) can only be justified if the joint material is ductile enough for redistribution to occur, especially in long joints where the effect is more marked. British Standard BS.8118 caters for this by deducting a further amount ΔL from the effective length (beside that needed for end defects) when $L > 10g$. Here ΔL is given by:

$$\Delta L = \frac{(0.1L/g - 1)L}{8} \tag{11.13}$$

where g is the throat dimension.

When a weld is under a clearly defined stress gradient, as in the web-to-flange connection in a beam, a conventional calculation can be used to obtain the load intensity.

Another case of a joint subject to non-uniform loading is a fillet group transmitting an eccentric in-plane force P (Figure 11.6). The procedure here is to resolve P into a parallel force P through the centroid G of the group and a moment M. At any given point in the weld, these produce parallel and tangential components of force intensity ($\overline{P}_1, \overline{P}_2$) given by:

$$\overline{P}_1 = \frac{P}{L_e} \tag{11.14a}$$

$$\overline{P}_2 = M\left(\frac{r}{\int r^2\,\mathrm{d}s}\right) \tag{11.14b}$$

where L_e is the total effective length of weld and r the distance of the point

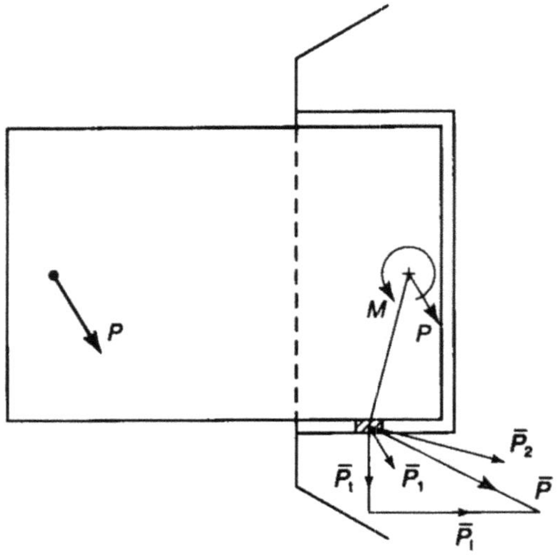

Figure 11.6 Fillet welded joint under eccentric load.

considered from G, the integration being performed over the full length of the joint. Components $\overline{P}_1$ and $\overline{P}_2$ are then combined vectorially to produce $\overline{P}$, and hence the required components $\overline{P}_t$ (transverse) and $\overline{P}_l$ (longitudinal).

11.3.4 Calculated resistance, weld-metal failure

The calculated resistance $\overline{P}_c$ per unit length of weld to either transverse or longitudinal loading on its own (Figure 11.4(a), (b)), based on weld-metal failure, may be found from:

$$\overline{P}_c = kagp_w \tag{11.15}$$

where: p_w = limiting stress for weld metal,
g = weld throat dimension,
a = factor governed by weld geometry,
k = 1.0 for butts or 0.85 for fillets.

The limiting stress p_w depends on the choice of filler, and Table 11.6 lists the p_w-values given by BS.8118. If the filler alloy is not known at the design stage, the lowest listed value of p_w should be taken for the parent alloy concerned, corresponding to the weakest filler. When a stronger filler is used, there may be a greater risk of cracking and hence a need for tighter control in fabrication. This might for example apply when a 5183 filler (Al–Mg) is used for joining 6082 plates, with a 15% strength increase compared to one of the more tolerant 4xxx-type fillers (Al–Si). For a weld between dissimilar parent metals, the lower of the listed values for p_w should be taken.

Table 11.6 Limiting stress p_w (N/mm²) for MIG or TIG weld metal

Filler alloy		Parent alloy						
		3103 3105	5251	5154A	5083	6063	6061 6082	7020 7019
Al–Mn	3103	80	–	–	–	–	–	–
Al–Si	4043A	90	–	–	–	150	190	–
	4047A	–	–	–	–	150	190	–
Al–Mg	5154A	–	–	210	–	–	–	–
	5356, 5056A	–	200	215	245	155	205	255
	5183, 5556A	–	200	215	275	165	220	265

The weld throat dimension g is defined as the minimum dimension through the weld metal, measured from root to outer surface and ignoring any reinforcement or convexity (Figure 11.7). Any penetration into the preparation at a butt or at the root of a fillet would also be ignored, unless clearly specified.

The factor α depends on the angle ϕ between the line of action of the transmitted force and the assumed failure plane (the weld throat in this case). Figure 11.8(a) shows how ϕ is measured, while the curve in Figure 11.9 gives the variation of α with ϕ. The equation to this curve is:

$$\alpha = \frac{2 \sin \phi}{\sqrt{\{(1 - \cos 2\phi)^2 + 3 \sin^2 2\phi\}}} \tag{11.16}$$

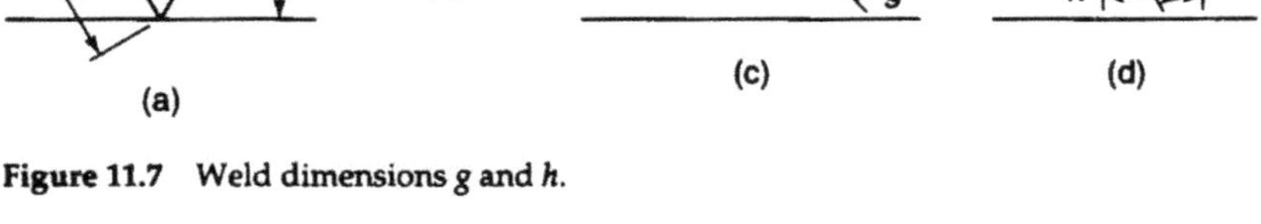

Figure 11.7 Weld dimensions g and h.

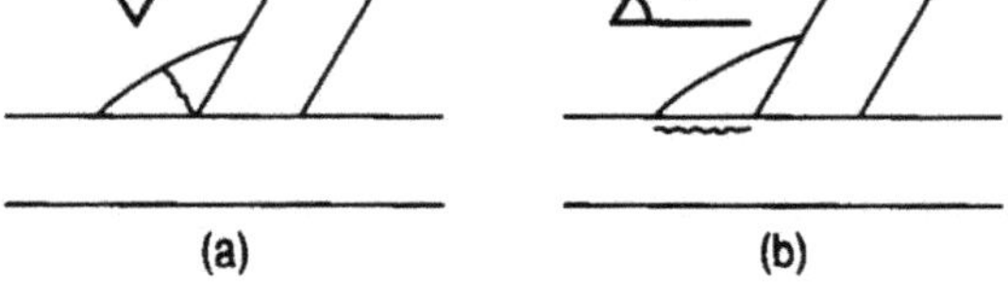

Figure 11.8 Definition of angle ϕ: (a) weld metal failure; (b) fusion boundary failure.

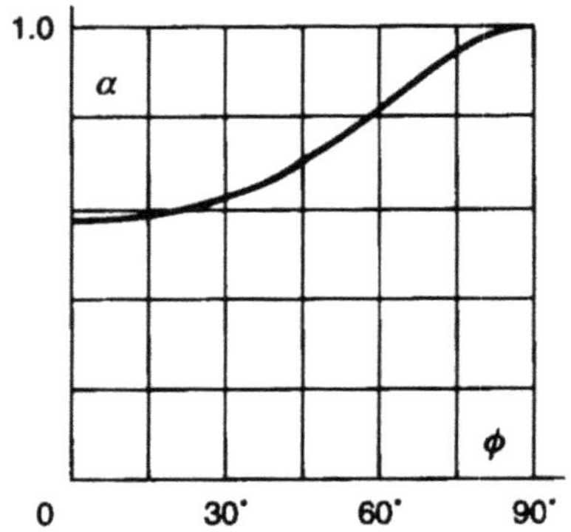

Figure 11.9 Variation of α with ϕ (equation (11.16)).

salient values being:

ϕ	α
$\phi = 90°$	$\alpha = 1.00$
60°	0.82
45°	0.71
30°	0.63
0	0.58.

Appropriate values for certain cases of weld metal failure are therefore:

Transverse in-line butt ($\phi = 90°$)	$\alpha = 1.00$
Transverse fillet lap-joint ($\phi = 45°$)	$\alpha = 0.71$
Any case of longitudinal loading ($\phi = 0$)	$\alpha = 0.58$

'Butt-fillet' welds (Figure 11.7(d)) cause difficulty, because it is not clear which value to take for k. When the preparation has been specified, but the width of the deposit is uncertain, a safe procedure is to take $k = 1.0$ and $g = g_1$ (Figure 11.7(d)). Alternatively, if the size of the deposit is clearly specified, it is acceptable to take $k = 0.85$ and $g = g_2$ should this prove more favourable.

11.3.5 Calculated resistance, fusion-boundary failure

At any weld there are two fusion-boundary planes, both of which may need to be considered. The calculated resistance $\bar{P}_c$ per unit length of weld is found from:

$$\bar{P}_c = \alpha h p_f \qquad (11.17)$$

where p_f = limiting stress for fusion-boundary failure, h = width of failure plane, and α = factor depending on joint geometry (as before).

The limiting stress p_f is governed by the properties of the parent metal. For heat-treated material it is given by:

$$p_f = k_{z1} p_a \qquad (11.18a)$$

where p_a = limiting stress for unwelded parent metal, and k_{z1} = HAZ softening factor (Section 6.4.1).

For the non-heat-treatable alloys it is found from:

$$p_f = 0.5(f_{oo} + f_{uo}) \tag{11.18b}$$

where f_{oo} and f_{uo} are the proof and ultimate of the parent metal in the annealed condition. Values of p_f thus obtained are listed in Table 5.4.

The width of failure plane h should normally be based on the nominal weld size (Figure 11.7), ignoring any penetration into the preparation at a butt or at the root of a fillet, unless this has been clearly specified.

As before, the factor α is a function of the angle ϕ between the line of action of the transmitted force and the assumed failure plane (Figure 11.8b). It may be read from Figure 11.9 or calculated from equation (11.16). For a weld under longitudinal loading, we again have $\phi = 0$ and $\alpha = 0.58$.

11.3.6 Welded joints carrying axial moment

A welded joint required to transmit moment about its longitudinal axis should be of one or other of the forms shown in Figure 11.10: (a) containing two parallel welds; or (b) full penetration butt-weld.

With (a), the two welds in the joint are effectively subjected to equal and opposite transverse forces as shown, and can be designed accordingly. With (b), the basic requirement to be satisfied at any position along the weld is that:

$$\overline{M} < \frac{\overline{M}_c}{\gamma_m} \tag{11.19}$$

where $\overline{M}$ = moment transmitted per unit length of weld when factored loading acts on the structure, $\overline{M}_c$ = calculated moment resistance per unit length of weld, and γ_m = material factor.

Here $\overline{M}_c$ should be taken as the lesser of two values determined as

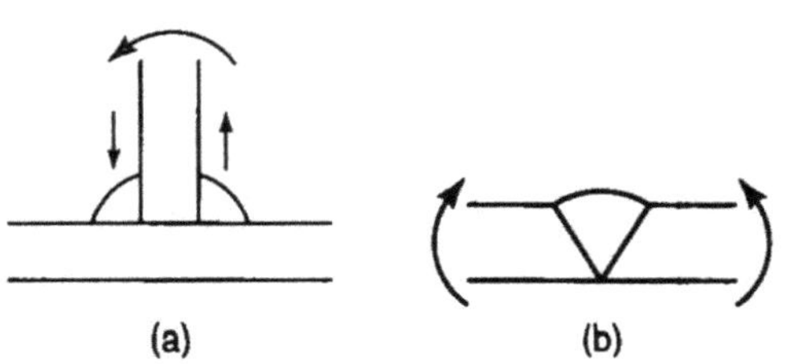

Figure 11.10 Welded joints transmitting axial moment.

follows, corresponding to an elastic stress pattern:

$$\text{Weld-metal failure} \qquad \overline{M}_c = \frac{g^2 p_w}{6} \tag{11.20a}$$

$$\text{Fusion-boundary failure} \qquad \overline{M}_c = \frac{t^2 p_f}{6} \tag{11.20b}$$

where p_w, p_f = limiting stresses (Sections 11.3.4, 11.3.5), g = weld throat dimension, and t = lesser thickness of the connected plates.

11.3.7 Welds under combined loading

(a) Transmitted force inclined to axis of weld

When, at any point in a weld, the transmitted force $\overline{P}$ is inclined at θ to the axis, the weld can be checked using an interaction diagram such as that shown in Figure 11.11,

where: $\overline{P}_t, \overline{P}_l$ = components of $\overline{P}$ acting transverse and parallel to the axis of the weld, under factored loading,
$\overline{P}_{ct}, \overline{P}_{cl}$ = calculated resistances per unit length to transverse and longitudinal force, each on its own,
γ_m = material factor.

The proposed interaction rule, using straight lines, is offered in preference to the British Standard rule (quadrant) because it avoids the need to worry about interaction when θ is close to 0° or 90°.

(b) Moment combined with transverse force

For a full penetration butt, having to transmit axial moment M at the same time as transverse force P_t, the requirement is that:

$$\overline{P}_t < \frac{x\overline{P}_{ct}}{\gamma_m} \tag{11.21}$$

where: $x = 1 - \gamma_m \overline{M}/\overline{M}_c$
$\overline{P}_t, \overline{M}$ = transmitted force and moment per unit length, when factored loading acts on the structure,
$\overline{P}_{ct}, \overline{M}_c$ = calculated resistances per unit length, each on its own,
γ_m = material factor.

(c) Moment combined with transverse and longitudinal force components

When a full penetration butt is required to transmit a longitudinal force, in addition to axial moment and transverse force, the interaction rule in Figure 11.11 may be used with $\overline{P}_{ct}$ replaced by xP_{ct}.

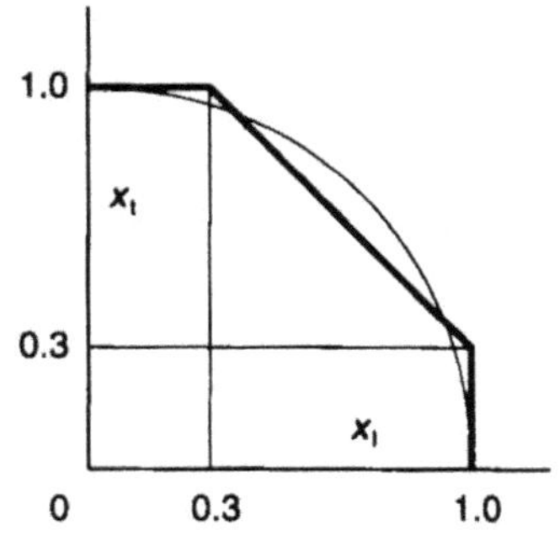

Figure 11.11 Interaction diagram for a weld carrying transverse and longitudinal components of load.
$x_l = \gamma_m \overline{P}_l / \overline{P}_{cl}$, $x_t = \gamma_m \overline{P}_t / \overline{P}_{ct}$

11.3.8 Friction-stir welds

The procedure for checking a friction-stir weld is simpler than that for arc welds (MIG, TIG), and at this early stage of FS development we very tentatively put forward the approach given below. The essential point is that there is no added weld metal to consider.

The basic requirement is to satisfy expression (11.11) or (11.19) as for an arc weld. In so doing, the calculated resistance per unit length of weld may be obtained as follows, depending on the kind of action effect transmitted:

Transverse force	$\overline{P}_c = tp_f$	(11.22)
Longitudinal force	$\overline{P}_c = 0.58tp_f$	(11.23)
Axial moment	$\overline{M}_c = \dfrac{t^2 p_f}{6}$	(11.24)

Here p_f is the same limiting stress that is used with fusion-boundary failure of arc welds (Section 11.3.5), while t is the thickness of the thinner connected part. Refer also to Section 6.9.

11.4 BONDED JOINTS

11.4.1 General description

An alternative to the riveting, bolting or welding of aluminium is to use glue. Fifty years of experience have shown this to be a sound procedure if properly done. The word 'glue', however, is not generally used and instead we refer to adhesive bonding or just bonding. The method is acceptable for use with all the alloy groups, and has the following advantages when compared with other connection methods:

1. absence of any weakening in the connected parts;
2. good appearance;
3. good fatigue performance;
4. good joint stiffness;
5. no distortion.

Unfortunately, there are some drawbacks in the use of bonding which tend to limit its range of application. Firstly, most adhesives lose strength at quite a modest temperature. Secondly, some will deteriorate when immersed in water (for a month or more). Thirdly, adhesives undergo creep under long-term sustained loading, if the stress level is too high. And, finally, there is a risk of cracking under impact conditions with some of them. All of these tendencies can be minimized by correct choice of adhesive.

Bonded joints are ideally designed to act in compression or shear, and the adhesives used with aluminium have a shear strength of around 20 N/mm^2. This relatively low value, which is only some 10% of that for the metal itself, can be accommodated by suitable design. One typical kind of application is when a built-up member is fabricated from two or more specially designed extrusions, employing adhesive in the longitudinal joints. Another is for lap joints between sheet-metal components. Bonding is not suitable for highly stressed joints, as at the nodes of a truss.

Bonded joints are inherently poor in their resistance to *peeling*, a form of failure that can occur when even a small component of tension, perpendicular to the plane of the adhesive, acts at the edge of a lap joint (Figure 11.12). It is essential to eliminate any risk of such a failure by intelligent design.

11.4.2 Specification of the adhesive

Adhesives used for bonding aluminium are either of the two-component or one-component type. With the former, curing begins as soon as the two components are combined, and then proceeds at room temperature, although it can be speeded up by modest heating. The one-component adhesives are cured by heating the glued component in an oven; they 'go off' much more quickly.

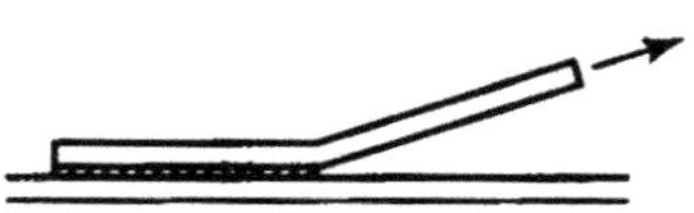

Figure 11.12 Peeling component of load on a bonded joint.

Manufacturers of structural adhesives offer a range of formulations of both types, designed to provide different properties for different applications. Unfortunately, there is no universal system of nomenclature in the way there is for aluminium materials, and each company markets its products using its own brand names. Below we provide data for a short-list of materials selected from the range produced by one particular firm, namely Ciba Speciality Chemicals of Duxford near Cambridge, UK, makers of the 'Araldite' adhesives. Equivalent materials are available from other suppliers.

11.4.3 Surface preparation

In making bonded connections, it is essential that the fabricator be meticulous in the preparation of the mating surfaces. The importance of this cannot be too highly emphasized. Three possible levels of surface treatment were defined in Section 3.6.2. It is up to the designer to specify the one to be used, and to ensure that it is properly carried out. The strength of the newly bonded joint is little affected by the choice of treatment. The point in specifying a more expensive one is to improve the reliability and durability of a joint, especially if it is to operate in a hostile environment or if its failure would be disastrous.

11.4.4 Effect of moisture

When a bonded joint is immersed in water over a long period, water becomes absorbed by the adhesive. This barely affects the strength of the adhesive itself, but may have a bad effect at the interface with the connected part. It must be appreciated that the adhesive sticks to the oxide layer and not to the actual aluminium. The integrity of the oxide and its adhesion to the underlying metal are affected by the surface treatment employed. Thus improved durability is obtained if the surface is abraded or grit-blasted before the final degrease, since this leads to a thinner and more tenacious oxide film than that on the aluminium as supplied. Better still is to use a chemical pretreatment or to anodize. Note, however, that the use of a more sophisticated surface treatment has negligible effect on the strength of the newly made joint.

Adhesive makers provide information about the drop in strength when a joint is immersed in water over a long period, and for some adhesives the fall-off can be considerable. This should not be interpreted as meaning that such adhesives are unsuitable for any outdoor use. It is only when a joint is actually immersed that the problem becomes significant, and this can usually be avoided. In most outdoor applications, the joint only gets wet occasionally, and the long-term loss of strength is relatively small. Bonded aluminium lighting columns, assembled after degreasing only, have performed well for 40 years.

11.4.5 Factors affecting choice of adhesive

In choosing the most suitable adhesive for a given application the following are possible factors to consider.

(a) Curing time/pot-life

For a mass-produced component, such as a car, the automatic choice would be for a one-component adhesive, because of a long available working time and the fast cure that is possible. With the two-component adhesives, the working time is shorter. Also the curing time is generally much longer, and varies greatly from one formulation to another. Here the choice is a compromise between the desire for a reasonable cure time and an acceptable *pot-life* (useable time after mixing).

(b) Toughening

Some adhesives are toughened. This means that they are specifically formulated so that after curing they contain an array of minute rubbery inclusions, which act as mini-crack-arresters. Such an adhesive is therefore more resistant to impact and would be selected when this is a factor. It will cost more.

(c) Slumping (loss of adhesive)

With a vertical joint, there is a danger that some of the adhesive will run out before it has had time to cure, leading to voids in the bonded area. This can be prevented by selecting an adhesive that does not flow in the precured condition, referred to as being *thixotropic*.

(d) Operating temperature

Many of the adhesives used on aluminium begin to lose strength at temperatures over 40°C. Some, however, are designed to operate up to 160°C.

(e) Performance in a wet environment

Some adhesives resist the effects of moisture better than others.

(f) Ductility

This can improve the strength of a joint by redistributing the load away from points of high stress. Most of the adhesives used with aluminium can be regarded as non-ductile in this respect.

11.4.6 Creep

Shear strength figures provided by adhesive manufacturers are normally based on short-term tests. When fully loaded over a long period, it is found that adhesives suffer from creep, characterized by continuing deformation and a drop in the shear strength. The effect becomes more pronounced as the temperature increases. Reliable data on this effect is not generally available, but a rough guide is that constant loading over a long period should not exceed about one-quarter of the short-term lap shear strength.

11.4.7 Peeling

Bonded lap joints are inherently vulnerable to premature failure by peeling, when subjected to loading perpendicular to the plane of the adhesive. This may occur even when the tension is only a secondary or accidental effect.

Manufacturers provide figures for the peel strength of their adhesives based on a standard test (the 'roller peel test'). This is expressed as the failure force per unit width of joint and typically ranges from 3 to 10 N/mm. An engineer should be encouraged to ignore such data and instead design joints so that peeling cannot occur. With joints between extrusions, this can be achieved by suitable design of the section. With lap joints between sheet-metal components, a possible technique is to employ occasional mechanical fasteners, strategically placed, to act as peel inhibitors, or else occasional spot welds can be used (through the adhesive).

11.4.8 Mechanical testing of adhesives

The most important mechanical property of an adhesive is its ultimate strength in shear. This is normally determined by testing a standard lap-joint specimen with the dimensions shown in Figure 11.13, composed of 2014A-T6 clad sheet. The shear strength τ of the adhesive is taken as the applied load at failure divided by the lap area.

Such a test is simple, but not ideal. Firstly, as the load builds up the

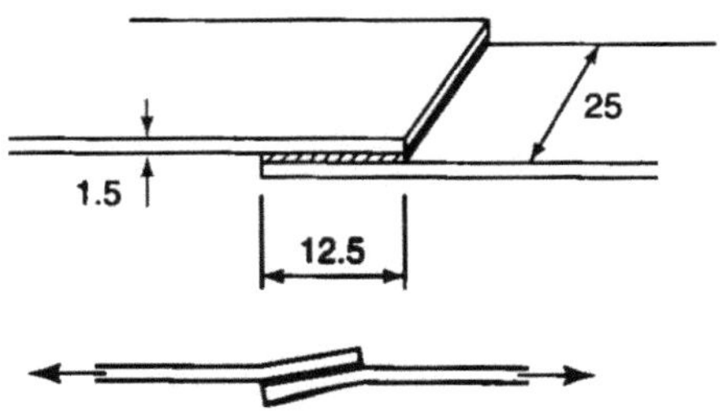

Figure 11.13 Adhesive lap shear test. Standard specimen.

aluminium bends slightly, as shown, with the result that a peeling component is introduced and the failure load thereby reduced. Secondly, depending on the relative stiffnesses of the aluminium (in tension) and the adhesive (in shear), there is a variation in the shear stress in the adhesive along the length of the joint, with a peak at either end. This produces a further drop in the failure load, because the adhesives used are not generally ductile enough to allow full redistribution of stress to occur, as would occur with a welded connection. The standard test is thus a convenient means for comparing different adhesives, but does not give the true intrinsic shear strength. It only provides a rough indication.

Possible tests for obtaining the intrinsic shear strength are the *thick adherend shear test* and the *butt torsion test* (Figure 11.14). Both eliminate the secondary effects inherent with the standard specimen. The ultimate shear stress obtained, using either, can be as much as 40% more than that found in the standard lap test.

Another kind of test employs a specimen composed entirely of adhesive, known as a *bulk tension specimen*, enabling the tensile properties of the adhesive material to be found. The ultimate tensile stress thus obtained is of fairly limited interest. But the value obtained for the Young's modulus E is useful, because it enables the shear modulus G to be deduced, the latter being needed when a joint is to be checked by finite element analysis. Such a test also provides the percentage elongation of the adhesive material at failure, which gives an indication of its ductility in shear, a significant factor in relation to redistribution of stress in a practical joint.

11.4.9 Glue-line thickness

Another factor affecting the performance of a bonded joint is the *glue-line thickness*, namely the thickness of the adhesive in the final joint as made. Ideally, for maximum strength, this should not exceed 0.3 mm for non-toughened and 0.6 mm for toughened adhesives. Precise data on the

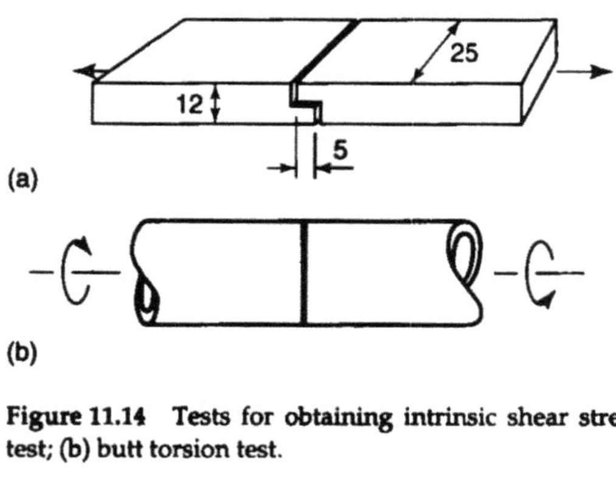

Figure 11.14 Tests for obtaining intrinsic shear strength of adhesive: (a) thick-adherend test; (b) butt torsion test.

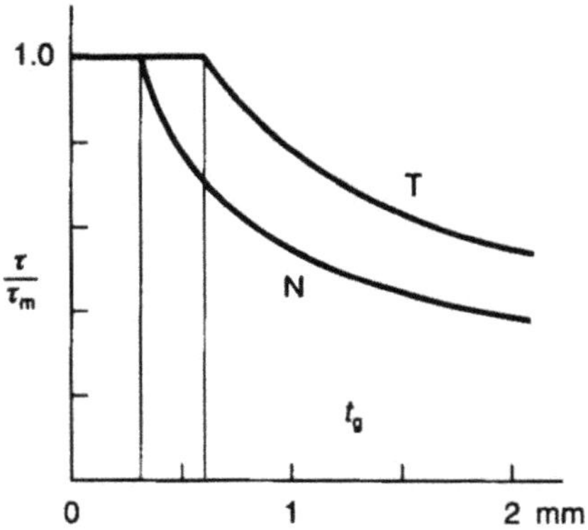

Figure 11.15 Effect of glue-line thickness on shear strength of adhesive. N = non-toughened, T = toughened.

fall-off in strength when these values are exceeded is not normally provided, but a rough rule is to assume (Figure 11.15):

$$t_g > t_{go} \qquad \tau' = \tau \sqrt{\frac{t_{go}}{t_g}} \tag{11.25}$$

where: τ' = shear strength with an over-thick glue-line,
τ = optimum shear strength ($t_g \leq t_{go}$),
t_g = glue-line thickness,
t_{go} = 0.3 mm for non-toughened adhesive,
= 0.6 mm for toughened adhesive.

In many situations, one can only control t_g very roughly, and its value may vary over the area of the joint, possibly reaching 1–2 mm in places. The designer should assume a liberal value in such cases. In a tongue-and-groove joint (Figure 11.16), it is theoretically possible to keep t_g down to the optimum value by suitable dimensioning, although care must be taken during assembly to ensure that the adhesive fully enters the groove. This can be assisted by widening the groove on the entry side to form a small 'reservoir'. Or else a taper can be introduced, in which case a low t_g can be achieved by in-plane clamping.

11.4.10 Properties of some selected adhesives

In the absence of universal standards we give the characteristics of seven selected epoxy type adhesives, taken from the 'Araldite' range made by Ciba Speciality Chemicals.

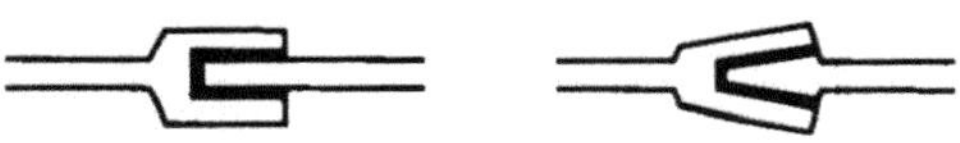

Figure 11.16 Tongue-and-groove bonded joints.

Table 11.7 Properties of six two-component epoxy adhesives

Araldite number	Pot-life (min)	Cure time at 23°C		Shear strength τ			Shear modulus at 23°C (N/mm²)	Special features
		τ = 1 N/mm²	τ = 10 N/mm²	at 23°C (N/mm²)	at 80°C (N/mm²)	after immersion (N/mm²)		
2010	8.5	30 m	3 h	23 (28)	2 (3)	23	?	Toughened, fast cure
2011	100	7 h	10 h	19 (28)	6 (8)	10	700	Ductile, best pot life
2012	4	20 m	1 h	17 (40)	4 (6)	7	?	Fast cure
2013	50–80	4 h	10 h	21 (22)	4 (5)	19	2500	Thixotropic, chem. resistant
2014	40	3.5 h	6 h	16 (18)	16 (18)	17	1800	Heat and chem. resistant
2015	30–40	4 h	10 h	21 (22)	7 (8)	14	1000	Toughened, thixotropic
1	2	3	4	5	6	7	8	9

Notes. 1 The table is based on data supplied by Ciba Speciality Chemicals of Duxford, Cambridge, UK, who manufacture the Araldite range of adhesives.

2. Pot-life is the useable time after mixing for 100 g of adhesive held at 25°C.

3. Columns 3 and 4 give the room temperature curing time to achieve $\tau = 1$ and $\tau = 10$ N/mm², when tested at 23°C.

4. Columns 5 and 6 give typical average strengths when tested at 23°C and 80°C, after curing as follows: first figure, 7 days/23°C; second figure (in brackets), 24 h/23°C + 30 min/80°C.

5. Column 7 refers to testing at 23°C after the specimen has been immersed in water at 23°C for 60 days. 16 h/40°C cure.

Table 11.7 and Figure 11.17 provide data on six selected two-component adhesives (Araldite 2010 to 2015). The quoted cure times to achieve a shear strength τ of 1 N/mm^2 indicate how soon an assembled component can be handled. Table 11.8 and Figure 11.18 give equivalent data for a useful one-component adhesive, namely Araldite AV119. All the quoted strengths, which are typical (not minima), are based on short-term tests using the standard lap specimen.

The information was provided by Ciba, the strength figures being unaffected by the alloy of the connected parts. It should be assumed that the given values refer to specimens having an ideal glue-line thickness, i.e. less than 0.3 and 0.6 mm respectively for the non-toughened and the toughened adhesives (Section 11.4.9).

Manufacturers are usually able to provide values of percentage elongation to fracture, as obtained from a bulk tension type of specimen (Section 11.4.8). For all but one of the listed adhesives, the elongation is low (1–4%). The exception is Araldite 2011, which is reported as reaching a strain of nearly 5% at maximum tensile stress, with 20% elongation at fracture (at a reduced stress). With 2011, therefore, there will be redistribution of the stress in a bonded shear-joint before final collapse occurs.

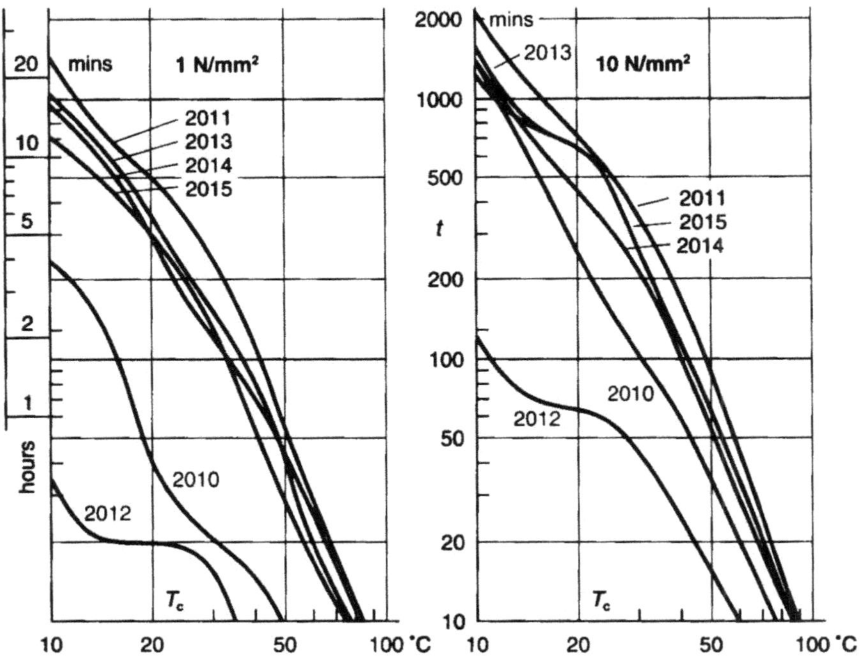

Figure 11.17 Curing times (t) for six selected two-component Araldite adhesives, to achieve shear strengths of 1 and 10 N/mm^2. T_c = curing temperature.

Table 11.8 Properties of a selected one-component epoxy adhesive

Araldite AV119		
Toughened		
Thixotropic		
Heat resistant		
Chemical resistant		
Typical short-term shear strength τ after 30 min/150°C cure:		
Tested at:	0°C	25 N/mm^2
	23°C	28.5 N/mm^2
	40°C	29.5 N/mm^2
	80°C	27 N/mm^2
	120°C	17.5 N/mm^2
Tested at 23°C after immersion in water at 23°C for 90 days		21 N/mm^2
Shear modulus (G) at 23°C		1300 N/mm^2

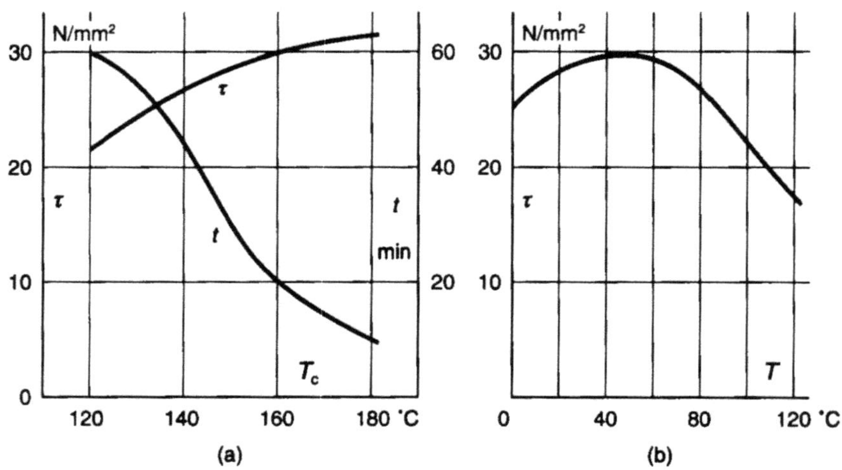

Figure 11.18 One-component Araldite adhesive AV.119: (a) curing time t and shear strength τ (at 23°C) as functions of the curing temperature T_c; (b) variation of shear strength τ with ambient temperature T after 30 min cure at 150°C.

11.4.11 Resistance calculations for bonded joints

Resistance calculations are needed in the earlier stages of design, even when a bonded joint is going to be eventually validated by testing. Such calculations pose problems because:

1. The stress transmitted by the adhesive is non-uniform, with peak values at critical points.
2. The adhesives used with aluminium are not generally ductile enough to allow redistribution of these peak stresses.
3. The shear strengths quoted for adhesives are only typical. Also they are based on the standard lap-test which is not ideal (Section 11.4.6).

When a joint is to be checked by calculation, the basic requirement is that the peak shear stress q_1 transmitted by the adhesive, under factored loading, should satisfy:

$$q_1 < \frac{p_v}{\gamma_m} \tag{11.26}$$

where p_v is the limiting shear stress for the adhesive and γ_m the material factor. The determination of suitable values for the three quantities involved is considered below.

(a) Stress arising (q_1)

The load transmitted by a bonded joint can be transverse or longitudinal, or a combination thereof. In each case, the adhesive is stressed in shear. The mean value q of the shear stress in the adhesive can usually be found using conventional engineering formulae. But this is not the same as the peak value q_1 which is what matters. Figure 11.19(a) shows the typical stress variation in a simple lap joint with peak stresses at the ends. For a simple joint of this kind, the difference between q_1 and q is considerable and not to be ignored, since the adhesives used are not generally ductile enough to allow full redistribution to occur. One outcome of this is that little advantage is gained by increasing the length l of such a joint beyond a certain point.

The situation can be improved by tapering the aluminium, which reduces the peakiness of the stress pattern (Figure 11.19(b)). Modifying the joint in such a way can reduce the peak stress to a value much closer to the mean, provided the taper is correctly designed.

The determination of q_1 and the design of the taper (if used) is typically achieved by employing an elastic finite-element program, although a simpler analysis is possible in some cases. An essential input to such calculations is the shear-flexibility of the adhesive, which is a function of its shear modulus G and the glue-line thickness t_g. In assuming a suitable value for the latter, it is important to realize that the 'peakiness' of the

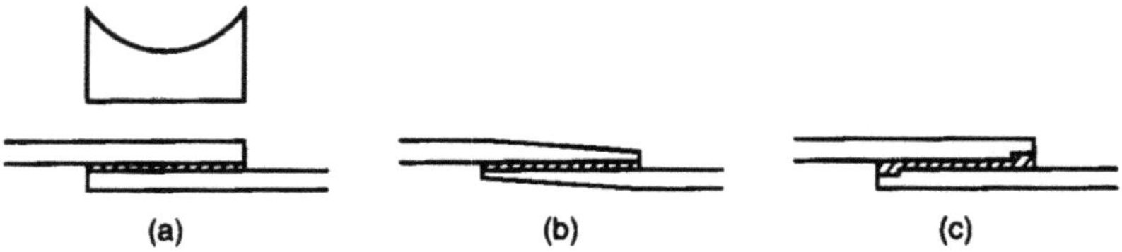

Figure 11.19 (a) Peaking of the stress at the ends of a conventional bonded lap-joint. (b), (c) Methods for making the shear stress more even.

stress pattern increases with the rigidity of the adhesive layer. Therefore, although increasing t_g above the optimum value decreases the actual intrinsic strength of the adhesive, this will partly be offset by the attainment of a less peaky stress pattern, having a more favourable value for the ratio q_1/q. The need for strict control of glue-line thickness thus tends to be less critical than might at first appear.

Figure 11.19(c) shows another method of reducing the peakiness, as used by the British aluminium designer Ron Cobden. The metal is cut back at each edge of the bonded area, as shown, thereby increasing the thickness of the adhesive and making it more flexible in shear (at the critical points). This device and also the tapering method are easy to incorporate with extrusions.

(b) Limiting stress (p_v)

An appropriate value to take for the limiting shear stress p_v for the adhesive depends on many factors and is difficult to arrive at. One possibility is simply to accept the value τ provided by the maker, based on the standard lap test. This is often a reasonable approach. Admittedly, the maker's listed value is only typical and not a guaranteed minimum. But against this, the standard test is clearly pessimistic and produces a strength value for the adhesive well below its intrinsic value. The second effect can be expected to more than compensate for the first, and the value thus obtained for p_v will probably be a safe one. Alternatively, for a more sophisticated assessment, the designer might call for special tests to be conducted to obtain the real intrinsic strength of the adhesive, using one of the specimen types in Figure 11.14.

Whatever the method used, it is obviously essential to allow for possible factors arising in service that might degrade the strength of the adhesive, such as: (a) operation at elevated temperature; (b) immersion in water; and (c) sustained loading (creep).

(c) Load factor

British Standard BS.8118 takes $\gamma_m = 3$ for bonded joints, this being the value called for when a joint is validated by testing. This relatively high value reflects the various uncertainties with the bonding process. The same value would be reasonable when the resistance is found by calculation.

11.4.12 Testing of prototype joints

It is recommended that the final design of a bonded joint should if possible be validated by means of tests, using specimens with the same construction and fabrication technique as for the final job. Such tests will lead to a value for the factored resistance, after dividing by γ_m, which can then be compared with the force it would be called on to transmit under factored loading.

British Standard BS.8118 requires a minimum of five tests, from which the factored resistance P_f is determined as follows:

$$P_f = \frac{P_m - 2S_d}{\gamma_m} \tag{11.27}$$

where P_m = mean failing load from all the tests, S_d = standard deviation, and γ_m = material factor, taken as 3.0.

CHAPTER 12

Fatigue

12.1 GENERAL DESCRIPTION

It is well known that seemingly ductile metal components can fail in a brittle manner at a low load, far below their static strength, when this load is applied many times. Aluminium is more prone to this problem than steel.

The phenomenon, known as *fatigue*, results from the presence of localized details or irregularities in zones carrying tensile stress, especially at welds. These act as stress-raisers and although they have no effect on static resistance, they become critical under repeated load. Elastic analysis predicts a peak stress at such positions that greatly exceeds the basic stress found using conventional stress formulae. The ratio of peak to basic stress, the stress-concentration factor, can reach a value of 3 or more. The peak stress, which is highly localized, causes a microscopic crack to form ('initiate') at a relatively low level of basic stress, which then grows ('propagates') each time the load is applied. At first the rate of propagation per load cycle is minute, but after many cycles it speeds up, eventually leading to catastrophic failure.

In non-welded construction, a fatigue crack may form at a bolt or rivet hole, at a sudden change of cross-section, or at any other geometric irregularity. Just the very slight surface roughness of the aluminium itself, well away from any joint or change of section, may be sufficient to cause fatigue. Welded components fare worse. Even when the welding is to the highest standard, there are still inevitable stress-raisers at the toe or root of a weld, and also in the ripples on the weld surface. These all lead to an inferior performance in fatigue. With lower standards of fabrication, the welds are likely to contain additional unintended defects (micro-cracks, undercut, lack of penetration), which will reduce the fatigue strength still further. The level of inspection specified to the fabricator can be crucial.

The number of cycles N to failure (the *endurance*) at a given detail is found to relate mainly to the *stress range* (f_r), especially for cracks initiating at welds. In other words, what matters is the difference between maximum and minimum stress in each cycle. Modern design rules for fatigue are therefore usually presented in terms of f_r, and ignore any slight influence that the mean stress might have.

Surprisingly, the choice of alloy has little effect on fatigue performance of members and the rules in current codes relate equally to all aluminium materials. Fatigue therefore becomes more critical with the stronger alloys, which are likely to operate at higher levels of stress in service.

The critical factor is the severity of the stress-raiser or defect. For example, if a cover plate is welded to the flange of an extruded beam, the stress range at the extreme fibres for a given fatigue life (say, 2 million cycles) may be reduced by 60%. Or, putting it another way, the anticipated life for a given range of extreme fibre stress might typically decrease by a factor of 30. Two other effects that can influence fatigue performance are:

- *Corrosion fatigue.* There is likely to be an added risk of fatigue failure if the structure has to operate in a very corrosive environment.
- *Scale effect.* For any given form of detail geometry, tests show that a thick component will be more prone to fatigue failure than a thin one.

A useful rough rule is to take the fatigue strength (limiting value of f_r) of an aluminium detail, for a given number of cycles, as one-third of that for a similar detail in steel. The fatigue data in BS.8118 aims to be more accurate than this and provides nine *endurance curves* of stress range f_r plotted against endurance N which are specific to aluminium. These are intended to cover most likely classes of detail, and are based on a large experimental programme using life-size specimens. These curves are generally more favourable than 'steel ÷ 3', especially at the high endurance/low f_r end.

The simplest situation is when the load cycles are of known and constant amplitude, as for a member supporting vibrating machinery. More often, there is a *load spectrum* comprising loads of varying amplitude and frequency, or even random loading. Often the most difficult problem in fatigue assessment is to estimate and then rationalize the pattern of the loading.

It is vital to identify the various types of loading that could lead to possible fatigue failure. These include:

- moving loads;
- vibration from machinery;
- inertia effects in moving structures;
- environmental loading (wind, wave);
- forces due to repeated pressurization;
- forces due to repeated temperature change.

There have been many instances of failure where the possibility of fatigue had not occurred to the designer. The author remembers the structure of a building in which a long aluminium tension member suffered failure even before the building had been clad. In service there was no possibility of fatigue, but the member had a low natural frequency of vibration and, in the unclad condition, wind-excited oscillations caused it to fail by flexural fatigue after a few weeks.

Another non-obvious type of fatigue failure is that due to transverse stressing at the welds in slender plate-girders. If the web operates in the post-buckled condition, due to a very high d/t ratio, it will flex in and out each time the load is applied, causing repeated flexure about the axis of the web/flange joint and hence fatigue in the weld.

The treatment of fatigue presented in this chapter is based on that in BS.8118, which was largely the work of Ogle and Maddox [30] at the TWI. The data provided for welded details refers specifically to arc-welded joints (MIG, TIG). Friction-stir welding is still in its infancy, but preliminary results suggest the FS process produces joints which are much better in fatigue than those made by MIG or TIG.

12.2 POSSIBLE WAYS OF HANDLING FATIGUE

There are three possible approaches for checking a proposed design against failure by fatigue:

1. safe life method;
2. fail-safe method ('damage-tolerant' approach);
3. testing.

The usual method (1), which is entirely done by calculation, is the one explained in this chapter. It essentially consists of estimating the range of stress f_r arising in service at any critical position, finding the corresponding endurance N from the relevant f_r–N curve, and then checking that the resulting life is not less than that required.

In method (2), the safety margins in design are lower than those required in a safe-life design. This is permissible because regular inspection is carried out, enabling the growth of any fatigue cracks to be monitored during the life of the structure. If the size of a crack or the rate of crack growth exceeds that allowed, the structure is taken out of service and the critical component repaired or replaced. Obviously, it is essential that all potential fatigue sites should be easily inspectable if this method is to be adopted, and considerable expertise is needed. Inspection methods, the time between inspections, acceptable crack lengths and allowable rates of crack growth must all be agreed between the designer and the user of the structure. When fatigue is critical, the fail-safe method will tend to produce a lighter structure than method (1). It is the approach most used in aircraft design. British Standard BS.8118 does not cover the fail-safe method, and it is beyond the scope of this book.

Fatigue testing (3) should be employed when it is impossible to apply method (1), due to problems in verifying a design by calculation alone.

For example:

- The loading spectrum is unknown and cannot be reliably calculated.
- The geometry of the structure makes stress-analysis difficult.
- It is not clear to which fatigue class a certain detail should be assigned.

Testing may also be preferred even when method (1) would be possible. For example with a mass-produced component, built to closely controlled standards of workmanship, it may be found that fatigue testing of prototypes would indicate a better performance than that predicted from the standard endurance curves. Advice on fatigue testing appears in BS.8118.

12.3 CHECKING PROCEDURE (SAFE LIFE)

12.3.1 Constant amplitude loading

The simplest type of fatigue calculation is when a single load is repeatedly applied to the structure, so that at any point there is a steady progression from minimum to maximum stress in each cycle without any intervening blips (Figure 12.1), referred to as *constant amplitude loading*. In such a case, the checking procedure at each potential fatigue site is as follows:

1. Decide on the design life of the structure. Refer to Section 12.3.3.
2. Calculate the number of load cycles n during the design life.
3. Determine the pattern and variation of nominal (unfactored) loading on the structure in each cycle.
4. Calculate the resulting stress range (f_r) at the position being considered – generally taken as the difference between maximum and minimum stress in each cycle. Refer to Sections 12.3.4 and 12.4.
5. Establish the class of the detail at the point considered. Refer to Section 12.5.
6. Using the endurance-curve appropriate to the class, read off the predicted number of cycles to failure (N) corresponding to the stress range f_r. Refer to Section 12.6.
7. The fatigue resistance at the point considered is acceptable if $N > n$.

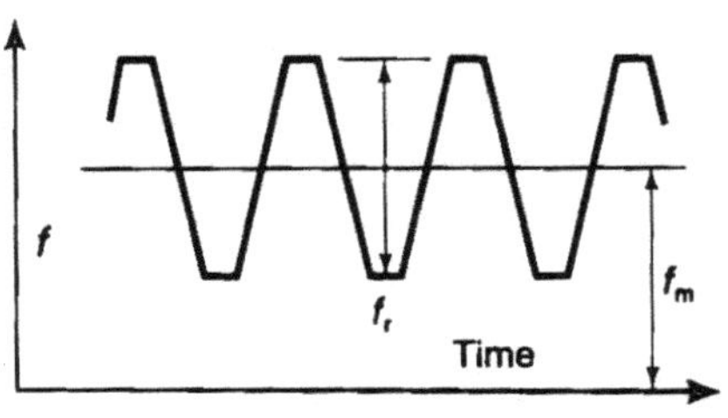

Figure 12.1 Constant amplitude loading. f_r = stress range, f_m = mean stress.

12.3.2 Variable amplitude loading

The simple state of affairs covered in Section 12.3.1 is fairly rare. In most fatigue situations, the loading is more complex, leading to a spectrum of stress ranges at any critical position. This is known as *variable amplitude loading* and the checking procedure runs as follows:

1. Decide on the design life of the structure, referring to Section 12.3.3 as before.
2. Find the number of load cycles during the design life.
3. Obtain the variation of nominal unfactored stress f in each cycle at the point considered (Figure 12.2). Refer to Sections 12.3.4 and 12.4.
4. Rationalize this stress history by reducing it to a set of specific stress ranges (f_{r1}, f_{r2}, f_{r3}, etc.), the number of times that each occurs during the design life being denoted by n_1, n_2, n_3, etc. This provides a *stress range spectrum* (Section 12.3.5).
5. Establish the class of the detail at the point considered. Refer to Section 12.5.
6. Select the appropriate endurance curve, and for each stress range value (f_{r1}, f_{r2}, f_{r3}, etc.) read off the corresponding endurance (N_1, N_2, N_3, etc.) that would be achieved if that stress range were the only one acting. Refer to Section 12.6.
7. The fatigue resistance at the point considered is acceptable if the Palmgren–Miner rule is satisfied:

$$\frac{n_1}{N_1} + \frac{n_2}{N_2} + \frac{n_3}{N_3} \cdots \nless 1.0 \tag{12.1}$$

12.3.3 Design life

The *nominal design life* of a structure is the time for which it is expected to be in service, and this should be agreed with the client. British Standard BS.8118 gives a range of typical values for a variety of applications.

The *design life*, as used in fatigue calculations, is normally taken the same as the nominal design life. However, the British Standard gives a designer the option of playing safer, if thought necessary, by multiplying the nominal life by a *fatigue life factor* γ_L (>1). A decision to do this would hinge on the accuracy of the assumed loading spectrum, whether records of loading will be kept, or the possibility of a change in use during the structure's life. It is fairly rare to step up the design life in this way.

12.3.4 Stress range

The stress range (f_r) is normally taken equal to the *nominal stress range*, namely the range over which f varies when nominal (unfactored) loads act on the structure. However, BS.8118 gives a designer the option to increase f_r by multiplying the nominal stress range by a factor γ_{mf} (>1). This might be felt advisable if: (a) the structure will have to operate in a very corrosive environment: or (b) failure at the position considered would result in total collapse, i.e. there is no alternative load path. In practise, it is fairly unusual to take $\gamma_{mf} > 1$.

British Standard BS.8118 allows a relaxation when f ranges from f_t tensile to f_c compressive, in which case the compressive component may be reduced by 40%. In other words, we then take $f_r = f_t + 0.6f_c$.

12.3.5 Stress-range spectrum

With variable amplitude loading, an essential step is to obtain the different stress ranges (f_{r1}, f_{r2}, etc.) in each cycle, and one possible procedure for so doing is the 'reservoir' method described in BS.8118. Referring to Figure 12.2, the graph showing the variation of f during the cycle is regarded as a reservoir, in which the greatest depth of water gives the value f_{r1}. The reservoir is then drained from its lowest point, the deepest remaining pocket (or pockets) giving the value f_{r2}. The process is repeated until all the water has been drained, thus obtaining f_{r3}, f_{r4}, etc. This enables a stress-range spectrum to be plotted, as shown in Figure 12.3.

This method is suitable when there is a sequence of loading events repeated many times. An alternative procedure is the 'rain-flow' method described in BS.5400: Part 10 (Steel, Concrete and Composite Bridges), which is more convenient when long and variable stress histories have to be analysed.

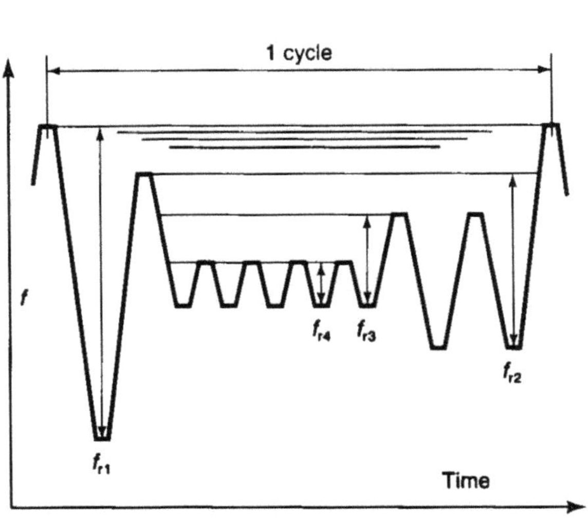

Figure 12.2 Variable amplitude loading, 'reservoir' method.

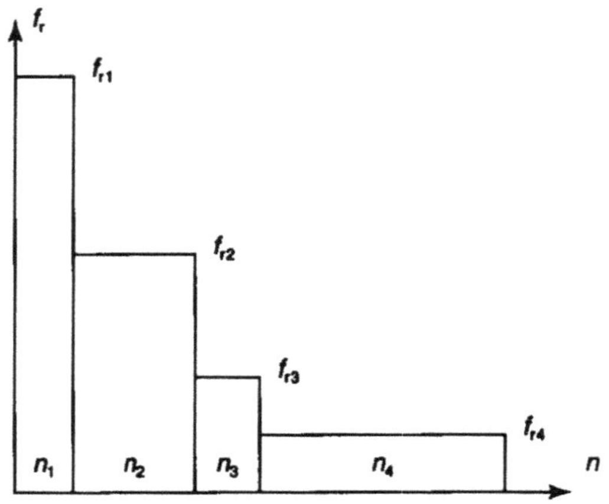

Figure 12.3 Stress-range spectrum.

12.4 REPRESENTATIVE STRESS

In determining the stress range (or stress-range spectrum) at a given fatigue site, it is important to know just what stress (f) we are talking about. There are essentially two methods (A, B) for defining f, the choice of which depends on the nature of the detail and the manner in which the crack propagates (Figure 12.4). Table 12.1 shows which method to use when.

12.4.1 Method A

In this method, f is taken as the major principal stress at the point of initiation, generally obtained by means of a simple analysis using conventional expressions such as P/A, My/I, etc., based on the gross cross-section without any reduction due to HAZ or local buckling effects. Local stress concentrations as at a small hole or the toe of a weld are ignored, this being justified by the use of a suitably lowered endurance curve that takes account of them.

Table 12.1 Choice of method for determining the representative stress f

Figure 12.4	Nature of the fatigue crack	Method
a	Crack at a non-welded detail	A
b	Crack which initiates at a weld and then propagates through parent metal	A
c	Cracks which initiate and propagate entirely within a weld:	
	(a) full penetration butt	A
	(b) partial penetration butt	B
	(c) fillet welds	B

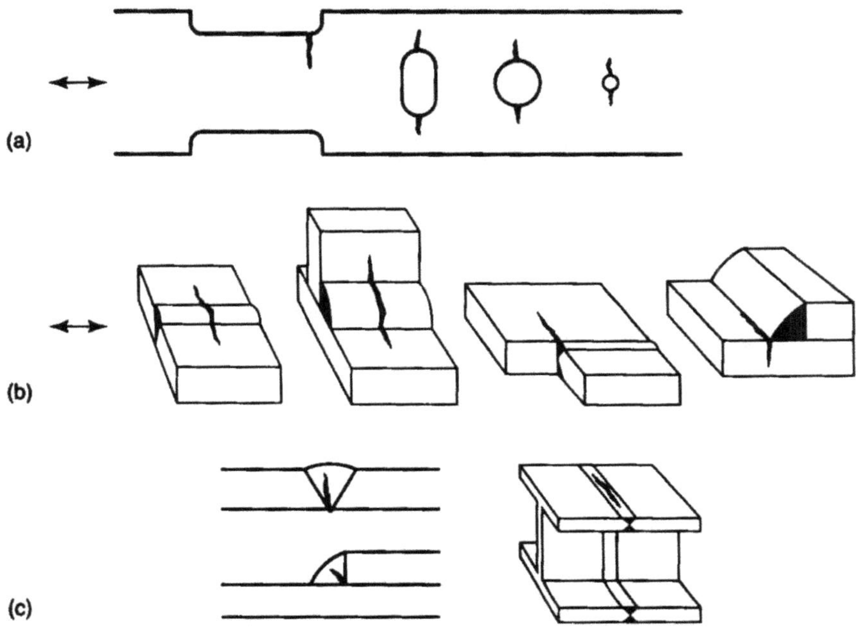

Figure 12.4 Crack propagation: (a) at non-welded details; (b) through parent metal at a weld; (c) through weld metal.

Larger geometrical effects receive a modified treatment whereby the basic stress is multiplied by a stress concentration factor K, enabling a higher endurance curve to be used. The factor K may be found from the literature, or else by means of a finite element analysis. For a member containing a large circular hole, we can generally take $K = 2.4$, while at a radiused change of section (Figure 12.5), K can be read from the curves

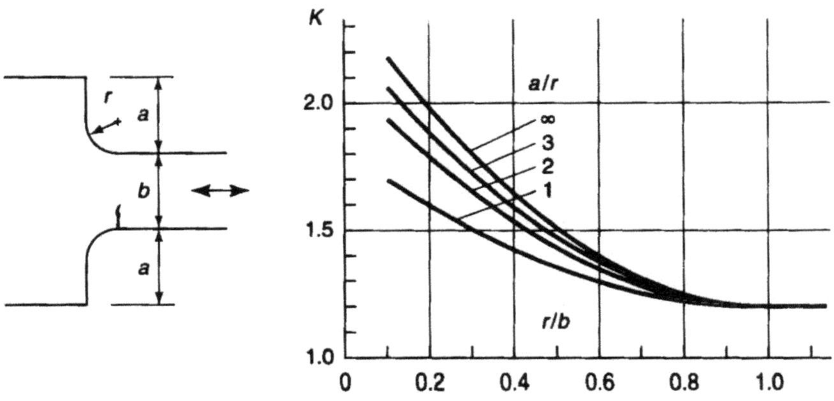

Figure 12.5 Stress concentration factor K at a radiused change of section.

provided, to which the equation is (valid for $a > r$):

$$\begin{aligned} &0.1 < r/b < 1 && K = 1.2\{1 + (1 - e^{-0.7a/r})(1 - r/b)^2\} \\ &r/b > 1 && K = 1.2 \end{aligned} \tag{12.2}$$

Other non-linear effects which become significant in fatigue are the secondary stresses in trusses, due to joint fixity, and the effects of shear lag, distortion and warping in plated structures. The increased stress levels resulting from these must be allowed for.

12.4.2 Method B

This is used for fillets and partial penetration butt welds transmitting load from one plate to another. A notional value is assumed for f obtained as follows:

$$f = \frac{\bar{F}}{ng} \tag{12.3}$$

where $\bar{F}$ = force transmitted per unit length of joint at the position considered, g = nominal throat dimension (Figure 11.7), and n = number of welds.

Here $\bar{F}$ can be a force transverse to the weld, a longitudinal one, or a vectorial sum of the two. It is normally found in the same general way as for $\bar{P}$ when considering static resistance (Section 11.3.3), except that we are now considering the force transmitted under nominal, and not factored loading.

When a single weld suffers bending about its longitudinal axis, f should be taken as the elastic flexural stress at the root, based on a linear stress distribution through the (nominal) throat. If necessary, this component of f should be added vectorially to the value found using equation (12.3).

Table 12.2 Classification of fatigue details (non-welded)

Case	Crack initiation position	Class
1	Extruded, rolled or drawn surface	60
2	Change of section	50
3	Cut out aperture	50
4	Large hole ($d > 3t$)	50
	Small hole ($d < 2t$):	
5	Open hole	35
6	For bolt securing minor attachment	35
7	Containing torqued HSFG bolt	29
8	Containing cold-driven rivet	29
9	Containing bolt in bearing	17

Notes. 1. Use K for cases 2, 3, 4.
2. An open hole having d/t in the range 2–3 may be treated as either case 4 (using actual stress concentration factor K), or case 5 (putting $K = 1$).

12.5 CLASSIFICATION OF DETAILS

12.5.1 The BS.8118 classification

An essential step in any fatigue calculation is to classify the form of the detail at the position being considered, so that the relevant endurance curve can be selected. British Standard BS.8118 distinguishes nine such classes, the reference number for each being the value of f_r (in N/mm^2) corresponding to a predicted endurance (N) of 2 million cycles. The class numbers thus defined are 60, 50, 42, 35, 29, 24, 20, 17, 14.

The class for a given detail may be found by referring to the relevant table, based on BS.8118:

Table 12.2 non-welded details;
Table 12.3 welded details, crack propagation through parent metal;
Table 12.4 welded details, crack propagation through the weld.

Table 12.3 Classification of fatigue details (arc-welded) – propagation through parent metal

Case	Crack initiation position	Class
	Longitudinal butt (away from weld end)	
10	Automatic, no stop/start	42
11	Manual	35
12	Dressed flush	50
	Longitudinal T-joint	
13	Automatic, no stop/start	42
14	Manual, away from weld end	35
15	At intermediate gap	29
16	At cope-hole	24
17	At end of connected item	20
	Longitudinal corner weld	
18	Automatic, no stop/start	42
19	Manual, away from weld end	35
20	At intermediate gap	29
	Longitudinal lap-joint	
21	Automatic, no stop/start	42
22	Manual, away from weld end	35
23	At intermediate gap	29
24	At end of connected item	17
25	As for 24, but double-sided	20
26	Through connected item	17
27	Transverse T-joint	29
	Transverse lap-joint	
28	$l < 25$ mm	29
29	$25 < l < 50$ mm	24
30	$l > 50$ mm	17
	At a local attachment	
31	$l < 25$ mm	29
32	$25 < l < 50$ mm	24
33	Abrupt side attachment	17

Notes. 1. For cases 26–30, avoid weld returns around lap.
2. l = length of connected part in direction of f_r.

Table 12.4 Classification of fatigue details (arc-welded) – propagation through weld

Case	Crack initiation position	Class
	Transverse butt, full penetration:	
34	Double-V prep, dressed flush	42
35	Double-V prep, as-welded	29, 35
36	Single-V prep, permanent backing bar	24
37	Single-V prep, unbacked	17
38	Butt-welded splice of a whole section, full penetration, dressed flush	24
39	Transverse butt, partial penetration	14
40	Root of fillet weld	14

Notes. 1. For cases 34–37 the ends of the weld must be ground flush, using run-on and run-off plates.
2. For case 35 the class depends on how closely the weld profile is controlled (i.e. preventing the use of excessive reinforcement).
3. For case 35 any change in thickness or width must be gradual, with tan $\theta \leq 0.25$.

For details not covered in these tables the reader may refer to the more comprehensive data provided in the British Standard.

The classes given in Tables 12.3 and 12.4 specifically refer to arc-welded joints made by MIG or TIG. They will only be valid if the fabrication meets a specified standard referred to in BS.8118 as 'fatigue-quality welding' (Section 12.7).

12.5.2 Friction-stir welds

At the time of writing, the new friction-stir process is being actively developed, and there are strong indications that FS welds will prove far superior in fatigue to those made by MIG or TIG. For example, a preliminary series of fatigue tests made by Hydro-Aluminium in Norway on transverse butt welds in 6082-T4 extruded material, 5 mm thick, suggest that class 50 might be appropriate for such a joint. This compares with class 24 for a single-V MIG weld with permanent backing bar, or class 17 if unbacked.

The reason for the better fatigue performance of FS welds is their flat as-welded profile. To obtain optimum results, it is important to avoid the small fin that can occur at the flow-zone edge of the nugget, when the welding parameters are not properly controlled. Provided this fin is eliminated, it is claimed that a fatigue class equal to 90% of that for the parent metal is possible.

12.5.3 Bonded joints

Bonded joints are generally superior to welded ones in fatigue. Consider first the performance of a member containing a longitudinal bonded joint, i.e. a joint extending in the same direction as the applied stress, as for the web/flange connection in a beam. For such a joint, it is reasonable to assume that there is little adverse effect due to the presence of the adhesive, and take a class approaching that for the basic wrought metal (class 60).

Secondly, consider the case when fatigue loading acts transverse to a bonded joint. Here there are two possibilities: tensile failure in the aluminium, and shear failure in the adhesive. The former can be handled in the same way as if the component were monolithic (no joint), taking due account of the stress concentration caused by the geometry. The latter is more difficult, due to lack of generally available data, and testing becomes necessary.

12.6 ENDURANCE CURVES

British Standard BS.8118 provides endurance curves in the form of stress range f_r plotted against endurance N (number of cycles to failure), covering the nine classes of fatigue detail. These are presented on a log–log plot, where they appear as a series of straight lines, the general form being shown diagramatically in Figure 12.6. Different curves are provided for constant and variable amplitude loading:

1. *Constant amplitude curves.* For any given class, line A is defined by the class number ($=f_r$ in N/mm^2 at $N=2\times10^6$) and the reverse gradient $1/m$ (as specified in Table 12.5). This line continues down to a horizontal cut-off at $N=10^7$.
2. *Variable amplitude curves.* For a given class, this uses the same line A, but only down to $N=5\times10^6$. There is then a transition line B of reverse gradient $1/(m+2)$ going from $N=5\times10^6$ to 10^8, followed by a (lower) horizontal cut-off.

For any given class, the equations to lines A and B are as follows, where the values of k_A and k_B appear in Table 12.5:

$$\text{Line A} \qquad f_r^m N = k_A \tag{12.4a}$$

$$\text{Line B} \qquad f_r^{m+2} N = k_B \tag{12.4b}$$

Also given in the table are the cut-off values (f_{oc} and f_{ov}) for each class, the idea being that a repeated stress range of lower magnitude is non-damaging. The variable amplitude case has a lower cut-off, because occasional stress ranges occurring above f_{oc} will produce some propagation, thus causing lower stress ranges to become damaging.

The resulting BS.8118 endurance curves are presented in Figure 12.7. Comparison of these with certain European data might suggest that at high N the British Standard endurance curves for welded details are too low. In fact, it would be wrong to conclude that the British Standard is

Table 12.5 Endurance curve parameters

Class	m	f_r at 5×10^6	k_A	k_B	f_{oc}	f_{ov}
60	4.5	48.9	2.01×10^{14}	4.82×10^{17}	42.0	30.9
50	4	39.8	1.25×10^{13}	1.98×10^{16}	33.4	24.1
42	3.5	32.3	9.60×10^{11}	1.00×10^{15}	26.5	18.7
35	3.25	26.4	2.09×10^{11}	1.46×10^{14}	21.3	14.9
29	3	21.4	4.88×10^{10}	2.23×10^{13}	17.0	11.7
24	3	17.7	2.76×10^{10}	8.63×10^{12}	14.0	9.7
20	3	14.7	1.60×10^{10}	3.48×10^{12}	11.7	8.1
17	3	12.5	9.83×10^{9}	1.54×10^{12}	9.9	6.9
14	3	10.3	5.49×10^{9}	5.85×10^{11}	8.2	5.7

Note. Stresses (f) are in N/mm^2.

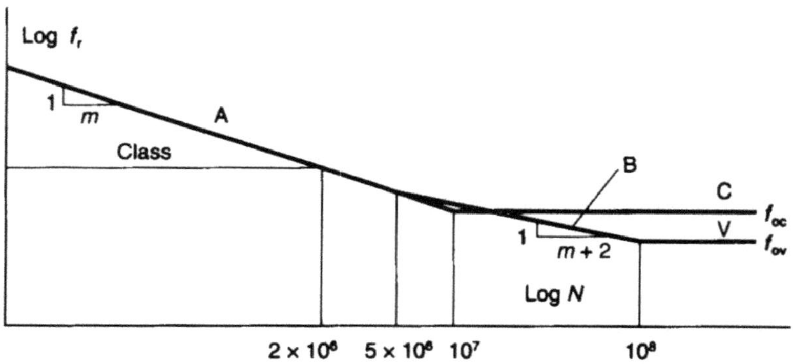

Figure 12.6 Construction of the endurance curves. C = constant amplitude loading, V = variable amplitude.

wrong, since the height of the curves in this range is critically affected by the standard of welding. The BS.8118 clearly states that the required quality of production welds must be easily achievable and assurable.

12.7 INSTRUCTIONS TO FABRICATOR

In order that the class of a welded detail (as given in Table 12.3 or 12.4) may be valid, the designer must specify fatigue-quality welding and state the necessary level of inspection (Section 3.3.5). The drawings should be marked with a 'Fat-number', giving the required class, and also an arrow showing the direction of the stress fluctuation.

The predicted endurance of a welded detail based on its normal classification will sometimes greatly exceed that actually needed, as when the design life is low or when a detail occurs in a zone of low stress. For such a detail, it is possible to calculate the *required class*, defined as the class of the lowest endurance curve that would produce the required life. By specifying this class instead of the normal (maximum possible) one for the detail considered, it becomes acceptable to relax the inspection requirements and hence save money. British Standard BS.8118 suggests the following procedure:

1. Determine those regions of the structure where the required class is equal to or higher than class 24.
2. At all details in such regions, the Fat-number shown on the drawing should indicate the required class, rather than the maximum possible class.
3. The fabricator then tailors his level of inspection to the Fat-number actually indicated.

An obvious example of where it is pointless to put the maximum possible Fat-number on the drawing is when a potentially high-class detail occurs adjacent to one of lower class. In such a case, the high-class detail may be marked with the same Fat-number as that for the low-class one next door to it, with nothing lost.

12.8 IMPROVEMENT MEASURES

If a critical detail fails to satisfy its fatigue check, the designer has essentially two options. One is to increase the section, thus reducing the level of stress. This increases weight and cost, and may be highly inconvenient if the design is far advanced. The alternative is to carry out an improvement measure, thus raising the fatigue class of the detail. The following are some of the possibilities:

1. *Redesign the detail*. The original low-class detail is replaced by one of higher class.
2. *Weld-toe grinding*. Grinding the toe of a transverse weld to a smooth profile reduces the stress-concentration effect. If the crack is toe initiated, the resistance to fatigue will be enhanced.
3. *Weld-toe peening*. Peening at the toe of a transverse weld introduces compressive residual stress, which again improves the performance when initiation is at the toe.
4. *Cold expansion of a bolt hole*, using a suitable drift. Like (3), this is helpful because it produces compressive residual stress at the point needed.
5. *Static overload*. Another way of introducing beneficial compressive residual stresses at potential fatigue sites is to give the whole component a controlled static overload to a stress beyond the elastic limit.

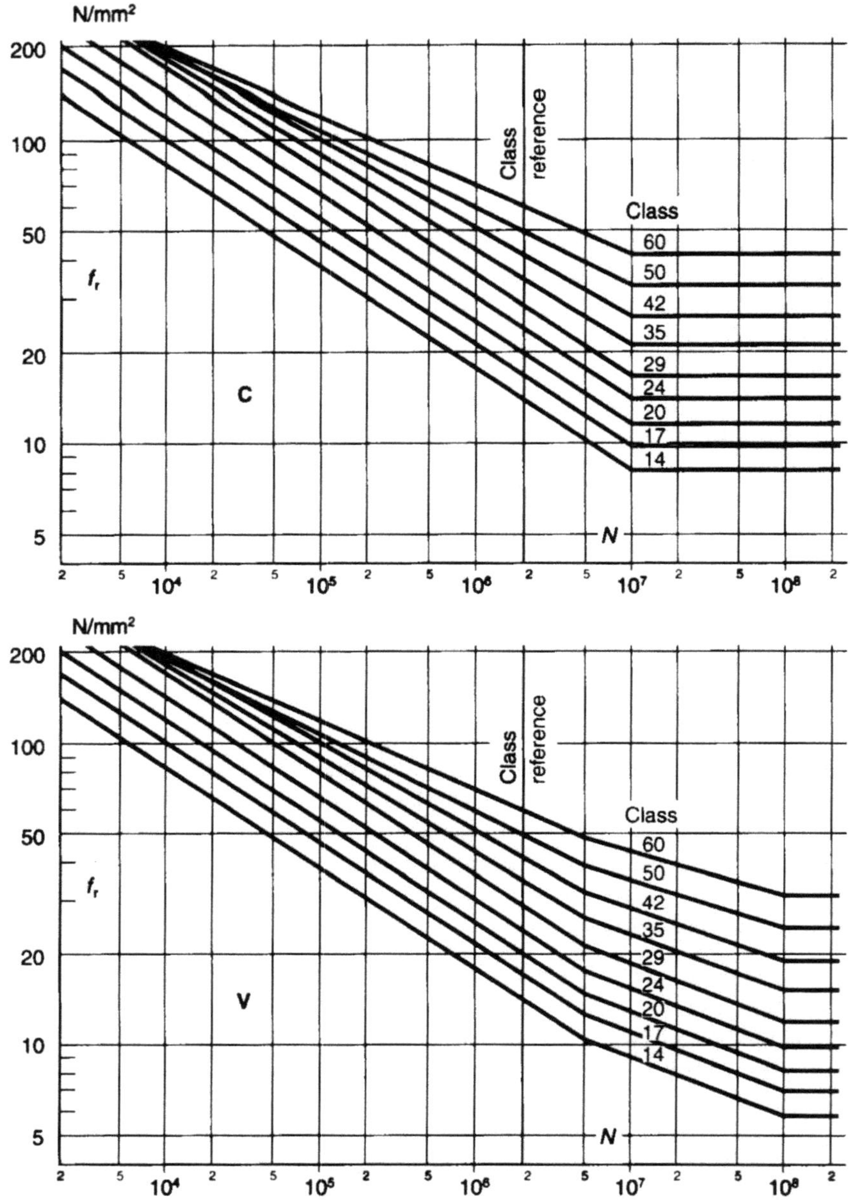

Figure 12.7 Fatigue endurance curves for aluminium. C = constant amplitude loading, V = variable amplitude.

Methods (2) and (3) are difficult to control, and one would normally resort to them only if fatigue problems had come to light during fabrication or even in service. Method (4) probably falls in the same category, although it is more controllable. Method (5) is a more reliable technique, provided the application of the overload is properly specified, and can be incorporated into the design. Methods (2) to (5) all require validation by testing.

12.9 FATIGUE OF BOLTS

12.9.1 Basic approach

Fatigue checks should also be made on any bolts the structure may contain, if these have to carry repeated tension. The root of a thread acts as a severe stress-raiser, causing bolts to perform poorly in fatigue. The first rule when fatigue is a factor is to avoid using bolts made of aluminium, because of the metal's intrinsic weakness in fatigue. The following approach therefore focuses on steel bolts, for which the checking procedure is the same as that used for checking a member, except in two aspects, namely, the appropriate endurance curve and the determination of the stress range. British Standard BS.8118 fails to give guidance on the fatigue of bolts used in aluminium structures.

12.9.2 Endurance curves for steel bolts

Aluminium designers can make use of the fatigue data for steel bolts provided in Part 10 of BS.5400 (Steel, Concrete and Composite Bridges). This may be expressed in the form of conventional endurance curves, as shown in Figure 12.8, the equations to which are as follows:

$$N < 10^7 \qquad N\left(\frac{f_r}{f_u}\right)^4 = 121 \tag{12.5a}$$

$$N > 10^7 \qquad N\left(\frac{f_r}{f_u}\right)^6 = 0.421 \tag{12.5b}$$

where: N = number of cycles to failure (with constant amplitude loading), f_u = ultimate strength of bolt material, and f_r = range through which the stress (f) varies.

The curves are radically different from those for members in that the strength of the material is now a factor. They are valid for steel bolts complying with BS.4395 (HSFG) or BS.3692 (Precision Bolts). They may also be used for black bolts (BS.4190), provided these are faced under the head and turned on the shank.

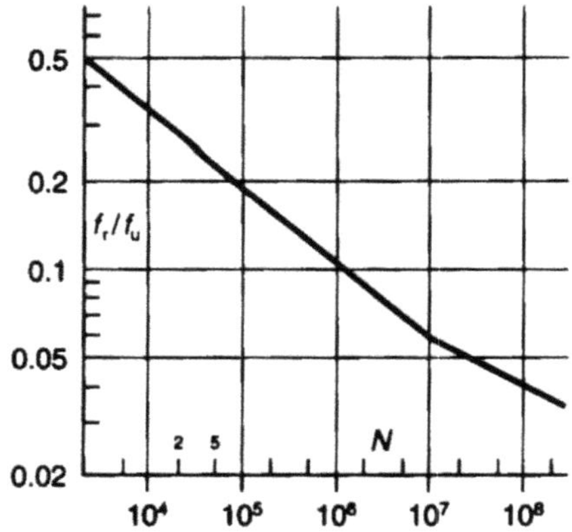

Figure 12.8 Endurance curves for steel bolts.

When fatigue needs to be checked for stainless steel bolts, used in an aluminium structure, it is suggested that the same endurance curves may be employed as those provided for ordinary steel ones, in view of the lack of readily available curves specific to stainless steel. This suggestion is believed to be valid if the bolts have rolled threads, but not necessarily with cut threads.

12.9.3 Variation of bolt tension

Figure 12.9 shows the typical variation in bolt tension T with the force T_1, defined as the applied force (per bolt) tending to pull the components apart. If the bolt were done up finger-tight, we would have $T = T_1$ all the way (line 1). If, however, the bolt is tightened to an initial preload T_0, the response will at first follow line 2, until separation occurs at C, after which

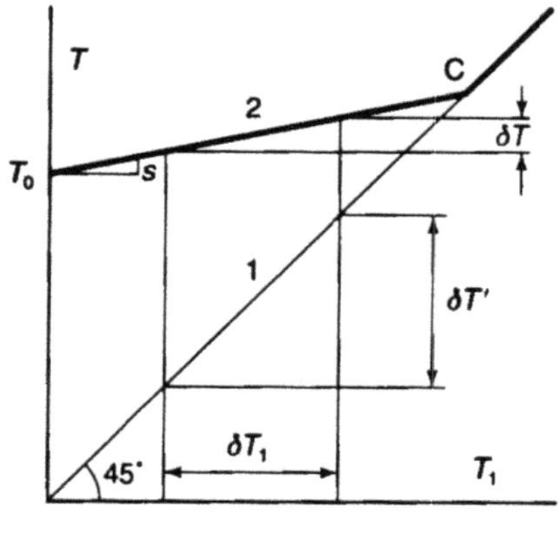

Figure 12.9 Variation of bolt tension T with applied external force T_1.

it reverts to line 1. The slope s of line 2 depends on the relative axial stiffnesses of the bolt and the clamped plate material. Fisher and Struik suggest that for all-steel construction s lies in the range 0.05–0.10. This tallies with a statement made in Part 10 of BS.5400: "The increase in tension will rarely exceed 10% of an external load".

When a steel bolt is used with aluminium plates, the slope of line 2 will be greater, because of the lower modulus of the plate material (one third). Extrapolating from Fisher and Struik's figures for steel/steel [33], a simple calculation suggests that s will now lie in the approximate range 0.14–0.25. Thus a reasonable upper-bound approach for the steel/aluminium case is to take:

$$\delta T = 0.25\, \delta T_1 \tag{12.6}$$

where δT_1 = change in T_1, δT = corresponding change in T.

It is suggested that the determination of the tensile stress range f_r for steel bolts used in aluminium structures should be based on δT as given by this equation. The fluctuation in bolt tension is seen to be more severe than for all-steel construction.

Figure 12.9 illustrates the importance of doing a bolt up tight when fatigue is a factor. If it were only tightened to a low initial tension, causing plate separation to occur in service, it would pick up the full fluctuation in the external force T_1 (shown as $\delta T'$ in the figure) instead of s times this. The range of variation in bolt tension would therefore be increased, possibly quadrupled, with a drastic effect on the fatigue life of the bolt.

References

1. *The Properties of Aluminium and its Alloys*, Aluminium Federation, Birmingham, UK, regularly updated.
2. Baker, J.F., and Roderick, J.W. (1948) The strength of light alloy struts, *Research Report No. 3*, Aluminium Development Association, London.
3. West, E.G. (1955) The first century of aluminium, *Bulletin of the Institution of Metallurgists*, March.
4. Gunston, W. (ed.) (1992) *Chronicle of Aviation*, J.L. Publishing, London.
5. Northern Aluminium Company, Development Department (1945) Light alloys in heavy engineering, *Light Metals*, February/March.
6. Craggs, H.R. (1958) Recent structural aluminium developments in Belgium, *Light Metals*, July/August, 213.
7. Spencer, H. (1989) *Aluminium Extrusions, A Technical Guide*, The Shapemakers (UK Aluminium Extruders Association, Birmingham).
8. Kissel, J.R. and Ferry, L.F. (1995) *Aluminium Structures*, Wiley, New York.
9. *Structural Use of Aluminium, Specification for Materials, Workmanship and Protection*, British Standard BS 8118: Part 2, 1991, British Standards Institute, London.
10. Thomas, W., Nicholas, D. and Dawes, C. (1994) Friction stir welding, *Metal Working Technology Europe*, 134–45.
11. Midling, O.T., Morley, E.J. and Kluken, A.O. (1994) Joining of aluminium constructions by friction stir welding, in *Proceedings of the International Symposium on Advanced Transportation Applications*, 3 November, Aachen, Germany.
12. *Registration Record of International Alloys Designations and Chemical Composition Limits for Wrought Aluminum and Wrought Aluminum Alloys*, The Aluminum Association, Washington, DC, USA, regularly updated.
13. Hepples, W. and Wale, D. (1992) *High Temperature Tensile Properties of 6082-T651*, Alcan International, Banbury Laboratory, Technical Assistance Report BTAR-93/024.
14. *Structural Use of Aluminium, Code of Practice for Design*, British Standard BS 8118: Part 1, 1991, British Standards Institute, London.
15. Bulson, P.S. (ed) (1992) *Aluminium Structural Analysis, Recent European Advances*, Elsevier Applied Science, London and New York.
16. Narayanan, R. (ed.) (1987) *Aluminium Structures: Advances, Design and Construction*, Elsevier Applied Science, London and New York.
17. Dwight, J.B. (1975) Use of the Perry formula to represent the European column curves, *Steel Construction*, **9**(1), 10.
18. Hill, H.N., Clark, J.W. and Brungraber, R.J. (1962) Design of welded aluminium structures, *Transactions of the American Society of Civil Engineers*, **127**(II).
19. Grong, O. and Myhr, O.R. (1993) Modelling of the strength distribution in the

heat affected zone of 6082-T6 aluminium weldments, in *Mathematical Modelling of Weld Phenomena* (eds H. Cerjak and K.E. Easterling), Institute of Metals, London.

20. Draft *Eurocode 9: Design of Aluminium Alloy Structures*, pre ENV 1999-1: 1997, European Committee for Standardisation, Brussels.
21. Robertson, I. and Dwight, J.B. (1985) HAZ softening in welded aluminium, in *Proceedings of the 3rd International Conference on Aluminium Weldments (INALCO)*, Munich, Germany, Aluminium-Verlag, Dusseldorf, Germany.
22. Dwight, J.B. and Robertson, I. (1998) Extent of the softened zone at MIG welds in 6xxx aluminium, in *Proceedings of the 7th International Conference on Aluminium Weldments (INALCO)*, April, 1998, Cambridge, UK, Abingdon Publishing, Cambridge.
23. Shercliff, H.R. (1988) Strength loss in the heat-affected zones of weld in non-heat-treatable aluminium alloys, *Report CUD/C-Mat/TR149*, Cambridge University Engineering Department, Cambridge, UK.
24. Mofflin, D.S. and Dwight, J.B. (1984) Buckling of aluminium plates in compression, in *Behaviour of Thin-walled Structures*, (eds J. Rhodes and J. Spence), Elsevier Applied Science, London and New York.
25. Bulson, P.S. (1970) *The Stability of Flat Plates*, Chatto and Windus, London.
26. Mazzolani, F.M. (1995) *Aluminium Alloy Structures*, 2nd edn, E&FN Spon, London.
27. Trahair, N.S. (1993) *Flexural-torsional Buckling of Structures*, E&FN Spon, London.
28. Palmer, P.J. (1953) The determination of torsion constants for bulbs and fillets by means of an electrical potential analyser, *Research Report No. 11*, Aluminium Development Association, London.
29. Heyman, J. (1982) *Elements of Structural Analysis*, Cambridge University Press, Cambridge.
30. Ogle, M.H. and Maddox, S.J. (1985) Design rules for fatigue of aluminium structures according to BS 8118: Part 1, in *Proceedings of the 5th International Conference on Aluminium Weldments (INALCO)*, April, 1992, Munich, Germany, Aluminium-Verlag, Düsseldorf, Germany.
31. Little, G.H. (1981) Collapse behaviour of aluminium plates, *International Journal of Mechanical Sciences*, **24**(1), 37–45.
32. *Durability of Aluminium in Building* (2nd edition, 1956), Northern Aluminium Co., Banbury (part of Alcan), booklet.
33. Fisher, J.W. and Struik, J.H.A. (1974) *Guide to design Criteria for Bolted and Riveted Joints*, Wiley, New York.

Index

For Product Safety Concerns and Information please contact our EU representative GPSR@taylorandfrancis.com
Taylor & Francis Verlag GmbH, Kaufingerstraße 24, 80331 München, Germany

www.ingramcontent.com/pod-product-compliance
Ingram Content Group UK Ltd.
Pitfield, Milton Keynes, MK11 3LW, UK
UKHW022011180926
475701UK00003B/315
* 9 7 8 0 3 6 7 8 6 6 2 9 7 *